The Forest That Fire Made

The Forest That Fire Made

An Introduction to the Longleaf Pine Forest

John McGuire, Carol Denhof, and Byron Levan

THE UNIVERSITY OF GEORGIA PRESS ATHENS

Athens, Georgia 30602
www.ugapress.org

Designed by Erin Kirk
Set in Minion
Printed and bound by Versa Press
The paper in this book meets the guidelines for permanence and durability of the Committee on Production Guidelines for Book Longevity of the Council on Library Resources.

Most University of Georgia Press titles are available from popular e-book vendors.

Printed in the United States of America
27 26 25 24 23 P 5 4 3 2 1

Library of Congress Cataloging-in-Publication Data
Names: McGuire, John R., author. | Denhof, Carol, 1971– author. | Levan, Byron, 1988– author.
Title: The forest that fire made : an introduction to the longleaf pine forest / John McGuire, Carol Denhof, and Byron Levan.
Description: Athens : The University of Georgia Press, [2023] | "A Wormsloe foundation nature book"—Title page verso. | Includes bibliographical references and index.
Identifiers: LCCN 2022047070 | ISBN 9780820363592 (paperback)
Subjects: LCSH: Longleaf pine—United States. | Fire ecology.
Classification: LCC SD397.P59 M34 2023 | DDC 634.9/751—dc23/eng/20221115
LC record available at https://lccn.loc.gov/2022047070

Contents

Foreword

Had this book been published three decades ago, few landowners, land managers, and members of the general public would have paid much attention. How times have changed! Thirty years ago, longleaf forests were scarce and greatly fragmented—essentially museum pieces. In addition, most people considered trees to be the only significant components of longleaf pine forests. Today, longleaf forests are a topic of conversation among the very public that knew or cared little about them only a few decades ago. Although the authors give trees special attention in this book (after all, we are talking about the longleaf forest), today's conception of longleaf forests, and forests in general, is the total ecosystem contained and operating within its confines. Words like "ecosystem" and "biodiversity," commonplace today, were added to our vocabulary fairly recently.

Thirty or so years ago, forestry and forest management functioned primarily to produce profitable timber, with few exceptions in federal lands, wildlife preserves, and recreational parcels. Typically, fast-growing tree species specifically bred for rapid growth, disease resistance, and wood quality were planted and nurtured until a relatively early harvest, then the cycle repeated. Forest management was dominated by forest industry and its drive to produce a valuable crop at the lowest cost in the shortest possible time. The industry did so profitably and was quick to share those advances with private nonindustrial landowners. In 1980, the forest industry owned roughly 20% of the forest land in the Southeast and private nonindustrial landowners owned a whopping 70% of the resource. Various federal and state entities owned and managed less than 10% of the region's forest resource. Although the proportions have shifted slightly in the interim, the portion owned by federal and state entities remains relatively constant at less than 10%.

Interest in the enigmatic and endemic longleaf forest, which once dominated nearly 90% of the upland South, was limited to a few academics, government researchers, and private landowners. Dean Gjerstad and I created the Longleaf Alliance in 1995 at Auburn University as a small start-up nonprofit advocacy organization that focused on the conservation, restoration, and study of the longleaf pine component of southeastern forests. Longtime advocates included the U.S. Forest Service, the U.S. Fish and Wildlife Service, the U.S. Department of Defense, the Nature Conservancy, the Joseph W. Jones Ecological Research Center at Ichuaway, the Belle W. Baruch Institute of Coastal Ecology and Forest Science, and Tall Timbers Research Station in Thomasville, Georgia, but these had a myriad of other responsibilities and interests, while the Longleaf Alliance had only one. Cooperation and communication among these groups was an essential element in what followed.

The longleaf restoration community was growing and it included avid foresters as well as various "-ologists"—herpetologists, entomologists, mammologists, mycologists, soil scientists, fire ecologists, wildlife biologists, plant ecologists, and others—and an array of private nonindustrial landowners. By bringing together this diverse group of people with various interest levels and skillsets, we worked to achieve a twofold mission: to keep what we had of the longleaf resource through conservation, and to establish new longleaf forests where they likely existed in the past. Since the majority of longleaf-appropriate sites occur on private lands in the Southeast, we knew that the participation of private nonindustrial landowners was critical. Having their involvement hinged on the development of consistently successful planting techniques and management protocols as well as the identification and even development of markets for the unique set of products available from longleaf forests. After considerable trial and error, members of the longleaf community have made significant advances in the development of quality and affordable seedlings, the innovation of container-grown seedlings, the extensive research and development of safe herbicide treatments, and research into planting techniques that have made reliable establishment and reestablishment of longleaf stands on appropriate soils and sites possible.

The economics of longleaf remains a challenge to private landowners contemplating an investment in longleaf forests. The argument for a quicker return on investment is bolstered by the relatively long intervals between commercial harvest of longleaf pine products and the relatively short duration of individual land ownership. Other forestry models frequently fit the needs of investors better than the needs of longleaf. As interest in longleaf grew, however, new markets emerged. Longleaf pine straw shortened the investment period and frequently returned more income to the landowner than the timber itself. New developing markets, like carbon sequestration, have real promise and require few inputs beyond time. Eventually, it became apparent that, for many landowners, the challenge of creating and managing a historic forest—complete with all the components that make it unique—was reward enough.

Acres of longleaf pine seedlings were established on nonproductive farmland early in the restoration process. Once seedling and planting techniques were in place, establishing trees on the site was generally successful. The rest of the "forest ecosystem" was less certain. We used to say "better is better"—that is, longleaf trees are better than cotton; longleaf and broomsedge is better than longleaf alone; longleaf, broomsedge, and partridge pea (and other native legumes) is better than the same stand without them; that that same stand with Northern Bobwhite, gopher tortoises, and wiregrass . . . , and so on, until the entire suite of species is present. That is probably not possible, at least in one human lifetime, but it remains an ultimate and admirable goal.

Today, we take it for granted that most people recognize that a longleaf forest is more than the trees. The trees are a necessary component but no more significant than the other plant and animal community living in that forest. We can have longleaf trees without that rich community, but probably not the community without the trees. The trees themselves enable fire, which, while surprising to many, is a significant component to the health of the larger community. Frequent fires occurred naturally in longleaf forests and were deliberately set by both Native Americans and early settlers for a variety of goals. The longleaf tree is tolerant of fire essentially throughout its life, and the flammable cast-off needles make an excellent fuel for ground fires. The plants and animals of the longleaf ecosystem evolved with frequent low-intensity fires and flourish when it is present. Astute managers have developed fire regimes tailored to produce habitat and conditions that favor both the trees and the whole community. Someone once said that taking fire out of the longleaf forest is akin to taking the rain out of the rainforest.

Finally, although the diversity and natural beauty of managed longleaf forests may not be as marketable as other more tangible products, the satisfaction that can be taken in their production, maintenance, and appreciation is beyond measure. This book is an admirable attempt by knowledgeable scientists who share an interest in and passion for longleaf forest ecosystems. I highly commend this book to virtually any audience with a desire to know about one of the nation's most diverse forests with unparalleled natural and historical significance.

—**Rhett Johnson,** Cofounder and Past President, the Longleaf Alliance

The Forest That Fire Made

Well-managed longleaf pine forests are often thought of as parks because of their widely spaced trees and carpet of grasses dotted with wildflowers.

CHAPTER 1

Introduction to the Longleaf Pine Forest

The sun casts a reddish hue through a column of smoke that curls above the twisted and flattened tops of ancient longleaf pine trees. The forest below hisses and crackles as a fire inches its way along a carpet of golden grasses. Somewhere among the grasses, a gopher tortoise (*Gopherus polyphemus*) lazily plucks a ripe blackberry from a low-hanging bush. With seemingly little care for the approaching fire, the tortoise leisurely makes its way to a burrow opening in the sandy ground and slides down into darkness and safety below. As the fire continues its march through the web of grasses, numerous arthropods flush to find shelter high along the rough bark of the surrounding longleaf pine (*Pinus palustris*) trees. With the flash of its belly, an Eastern Phoebe (*Sayornis phoebe*) flies in from seemingly empty space, grabs a fleeing grasshopper, and quickly melts back into the safety of the forest and out of sight of prowling Cooper's Hawks (*Accipiter cooperii*).

> "In 'pine barrens' most of the day, low, level sandy tracts; the pine wide apart, the sunny spaces between full of beautiful abounding grasses, liatris, long, wand-like solidago, saw palmettos, etc., covering the ground in garden style. Here I sauntered in delightful freedom, meeting none of the cat-clawed vines or shrubs of the alluvial bottoms."
> —John Muir, *A Thousand-Mile Walk to the Gulf*

This drama has unfolded in the piney woods landscape of the Deep South for thousands of years. By all accounts the domain of longleaf pine was vast. These forests were seen in nine states: Texas, Louisiana, Mississippi, Alabama, Florida, Georgia, South Carolina, North Carolina, and Virginia. With the exception of occasional rivers, swamps, and Native American agroforestry fields, this forest stretched as far as the eye could see. In 1835, an English traveler described the bulk of his trip by stagecoach from Natchez, Mississippi, to Alexandria, Louisiana, as going through "hilly, pitch (longleaf) pine woods." Today's traveler on that route would not encounter such scenery.

A Forest Legacy

Few can dispute the diversity and beauty of a well-managed longleaf pine forest. The first impression is a forest smelling of pitch with tall, widely spaced pine trees. The visitor is also struck by how a lack of eye-level trees and shrubs and a well-developed, grassy ground layer helps to create a parklike scene. Though often called pine barrens by early explorers and settlers, these forests were far from barren. In fact, this unique forest was home to thousands of unique plant and animal species.

Longleaf pine forests were perhaps the most extensive coniferous forests found at one time. The same forest type dominated by longleaf pine seen in eastern Texas could also be found stretching to central Florida and northward to Virginia (an area of roughly 140,000 square miles). However, other pine forests can co-occur in the range of longleaf pine. Some are dependent on fire and serve similar habitat functions in the Southeast. Longleaf pine gives way to shortleaf pine (*Pinus echinata*) forests to the west and north; slash pine (*Pinus elliottii*) forests to the extreme south; and loblolly pine (*Pinus taeda*) forests, where fire is naturally infrequent, artificially suppressed, or promoted by forestry practices for timber production. Collectively, these pine forests (along with a few other of lesser occurrence) are known as the southern pines.

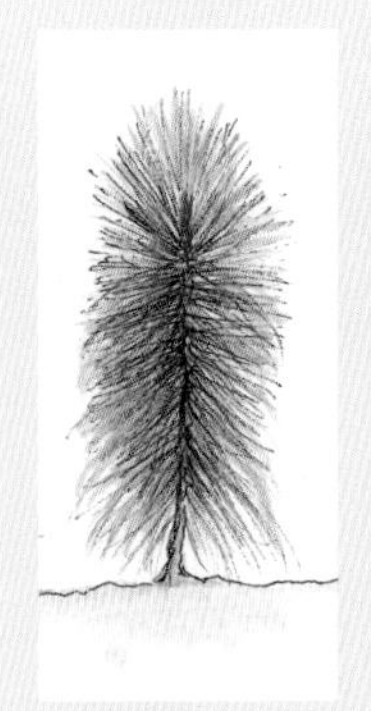

The original longleaf pine forest seen by early explorers to the southeastern United States shared several fundamental characteristics:

- Tall, majestic, and ancient stands dominated by a single species of tree—longleaf pine;
- A conspicuous lack of eye-level trees and shrubs presented a scenic vista through the forest;
- A well-developed ground layer, dominated by grasses, helped to create a parklike appearance;
- A high number of different plant types in the ground layer;
- Numerous wildlife species dependent on the open pine forest;
- Frequent fires of low intensity acted as the thread that held the longleaf pine forest together;
- Inhabiting a variety of soil types (sand to clay) and topographic (low and flat to mountainous) conditions.

Frequent, low-intensity fires moved through longleaf pine forests, helping to create the open and parklike setting.

Long-Lived Tree

Longleaf pine is the longest lived of the southern pine species. Throughout most of its range, individual longleaf pines have been documented up to 450 years old. Longleaf pine must go through several distinct growth stages to reach that age. Due to the large occurrence of patchy, small-scale disturbances, the forest as a whole is always simultaneously transitioning through the various stages of growth. An old-growth longleaf pine forest is a forest of large, scattered old trees with numerous pockets of younger trees.

Seed Stage. After falling from the tree between October and late November, winged seeds twirl to the forest floor and await adequate moisture before germination. In years of heavy seed production called mast years, a rain shortly after seed fall will yield a green blanket of sprouting baby longleaf pines on the forest floor. Any seeds that will germinate do so within a few weeks after falling. Although seeds will germinate almost anywhere (on rocks, logs, forest debris like pine straw), they generally need to land and become established on bare soil to survive subsequent periods with little rain. During this first stage of life, the seedlings are very susceptible to fire, drought, and predation from insects, such as ants, and small mammals, such as mice and squirrels. From here, it can take upward of a year to reach the next life stage.

Grass Stage. In this unique stage of a longleaf pine's life, the seedling resembles a clump of grass more than a tree. During the grass stage, the growing tip (bud) of the tree is protected under a thick arrangement of needles at ground level. This arrangement makes the tree resistant to fire.

When fires sweep through, the needles may burn, but the bud remains protected by the dense needles and silver scales. New needles quickly replace those that were burned off. Although the tree may not be growing upward during this stage, the seedlings put down an impressive root system. The grass stage typically lasts from 1 to 7 years, depending on the degree of competition with other plants for sun and moisture, although in rare instances this stage can last 20 years. Because of this distinctive ground-level growth form, animals are often found hiding beneath grass-stage seedlings.

The longleaf pine tree has distinct life stages from seed stage to after death. All stages provide value to the ecosystem and its inhabitants.

Bottlebrush Stage. When the diameter of the stem at ground level reaches 1 inch, the grass-stage longleaf will begin to initiate height growth. Often in March, a single, white growing tip will emerge upward from the protective sheath of needles. This white tip, called a candle, may grow a few feet in just a few weeks. By about late May, green needles begin to emerge from the candle, and parts of the candle turn scaly and brown as bark begins to form. At this point, the longleaf pine is growing proportionally more in height than it is in diameter. The tree often looks like a 3- to 4-foot-tall bottlebrush. By growing rapidly in a short period of time, the seedling is able to achieve enough height to gather sunlight and to get its growing tip above the char of frequent fires. However, during this stage of growth, longleaf pine trees are slightly more vulnerable to fire, as the tip is still relatively close to the fire's heat. It may be a year or more before the bark thickens enough to withstand most fires. The longleaf pine may remain in the bottlebrush stage for a few years. The thick bushy needles at this stage can be attractive to many insects trying to escape predators. Watch out: fiercely aggressive and stinging paper wasps (*Polistes carolina*) are among the insects that seem to prefer constructing their nests in seedlings this size!

Sapling Stage. When the longleaf reaches about 6–10 feet in height, lateral branches begin to emerge and signal the beginning of the sapling stage. Stem diameter increases and bark thickens modestly, but the tree continues to grow in height at upward of 3 feet per year. In the spring, white growing tips can be seen extending upward from the tufted needles at the end of the branches. As the tree grows taller and the bark thickens, the longleaf becomes less susceptible to fire. Once the tree reaches 8 feet in height and about 2 inches in diameter at ground level, it is rarely killed by fire. The tree will remain in this stage for several years. Side branches become attractive perches for many songbirds. Some birds such as the Pine Warbler (*Setophaga pinus*) may even begin to construct nests on the outer edges of these branches, concealed in the clumps of needles.

Mature Stage. Around 30 years after upward growth, longleaf pine trees begin to produce cones with fertile seeds. As the tree begins to mature, lower limbs may be shed or pruned by fire. The trunk begins to fill out into a straight, relatively branch-free tree that resembles a living telephone pole (many longleaf pines in fact wind up as telephone poles). On more fertile soils, the tree may continue to grow to heights exceeding 100 feet. On poorer soils, the tree may only grow to the height of a five-story building, or about 50 feet. After about 70 to 100 years, longleaf essentially stops growing in height. During the later stages of this period, trees may begin to show signs of decay and rot. In particular, longleaf pine reaching 80 years of age may become infected with the red heart fungus, which causes the otherwise dense heart of the tree to become punky, soft, sappy, and full of small channels. This spongy tree center is needed for the Red-cockaded Woodpecker (*Picoides borealis*) to construct its cavity.

Old Growth. Large-diameter trees with flat-topped crowns dominate this forest stage. Historical accounts describe longleaf pines over 120 feet tall and 3 feet in diameter. Some have suggested that old-growth longleaf pine trees stop growing at these advanced ages. However, there are many cases in which old-growth longleaf pine trees have actually increased growth rates at 200 years old or older when resources (light, soil moisture, nutrients) became available. As the forest ages, older trees begin to show signs of internal rot from red heart fungus. Dead spots such as those sites created by old broken limbs may form on the tree, and birds such as Pileated Woodpeckers (*Dryocopus pileatus*) may take up residence by constructing cavities. Bees have also been seen building their hives in these dead spots.

Death. In a landscape that regularly sees lightning, tornadoes, wildfires, drought, hurricanes, or even ice storms, it is remarkable for a longleaf pine to die from old age. After 300 and up to 450 years, trees that survive everything that nature has to throw at them will eventually weaken and begin to lose the ability to fend off forest pests like black turpentine beetles (*Dendroctonus terebrans*) or southern pine beetles (*Dendroctonus frontalis*). Slowly the trees die off. The initial signs of this weakening include a thinning of green needles in the tree top, followed by signs of beetle activity on the bark, then wilting of needles and finally by complete defoliation of the pine needles. Black turpentine beetles, southern pine beetles, and sculptured pine borers (*Chalcophora virginiensis*) all flourish as the tree's health declines. In response, birds such as Red-headed Woodpeckers (*Melanerpes erythrocephalus*) feast on these insects.

After Death. Longleaf pine continues to contribute to the ecosystem following its death. Dead standing trees are referred to as snags. Once the tree dies, its bark quickly sloughs off or is torn off by foraging woodpeckers. Sloughing bark provides habitat for many species, such as the scarlet kingsnake (*Lampropeltis triangulum elapsoides*). Secondary cavity nesters such as White-breasted Nuthatches (*Sitta carolinensis*) or flying squirrels (*Glaucomys volans*) may take up residence in these trees. What remains of the longleaf pine tree is its white skeleton. Snags that remain upright do so only for a few years, though there are instances of snags lasting much longer. Typically, without the protective bark in place, the spongy outer wood of the longleaf is exposed and often causes the snag to ignite during a forest fire and burn to the ground.

"What thrills me most about longleaf pine is how the pine trees sing. Even in the fragments you can hear music. Horizontal limbs of flattened crowns hold the wind as if they are vessels, singing bowls, and stir in them like a whistling kettle."
—Janisse Ray, *Ecology of a Cracker Childhood*

Anatomy of a Longleaf Pine

The Tree Top. Trees often stop growing upward at about 70 years of age or the height of a six- to eight-story building (60–90 feet). At this point, older trees begin to develop a characteristic flat top. The branches in these flattened tree tops begin to thicken over time and take on a crooked appearance. It's not unusual to see century-old or older trees missing half of their top, lost to lightning, wind, or ice storms. These flattened tops have the advantage of keeping individual trees from becoming substantially taller than the rest, which would make them lightning rods in a flammable forest.

Longleaf pines do not compete well with each other or with other tree species for limited resources. Most stands of longleaf pine consist almost exclusively of longleaf pines, and no other trees. Also, unlike many forest types, the crowns of longleaf pine are usually widely spaced, forming an open canopy that allows a great deal of sunlight to reach the forest floor. These open forests force animals like fox squirrels (*Sciurus niger*) to run down the tree, along the ground for several yards, and run up the next tree. No jumping from tree to tree.

Although the thin tops of longleaf pine generally do not provide much refuge from hawks, owls, and other winged predators, many animals and insects may seek temporary safety in the treetops when fire sweeps across the forest floor. Longleaf pine tops provide an excellent lookout for the sharp eyes of numerous predatory birds (e.g., the Southeastern Kestrel [*Falco sparverius paulus*]) scanning the forest floor for food such as lizards and small mice.

In an environment where lightning occurs frequently, there is little advantage to being the tallest tree. In response, the tree tops of longleaf pine trees flatten out over time.

Longleaf Pine Needles. Longleaf pine have the longest needles of all southern pines, a characteristic that gives them their name. The needles are grouped in threes and arranged in tufts on the end of branches. When viewed from below, these tufts look similar to a pompom. Needles persist for two to three growing seasons before being rotated out by a group of younger needles.

Compared with the other southern pines, longleaf pine needles are slow to decay once they fall to the ground. However, frequent fires throughout the woods easily ignite the flammable resins in the fallen needles and clean the forest floor.

Needles were commonly used by Native Americans and early settlers to weave into baskets. Today, because of their pleasing color and length, many longleaf pine needles are gathered and used as garden mulch (called pine straw) to control weeds.

The needles of a longleaf pine are (as the name implies) the longest of the southern pines. Their high resin content makes them attractive to landscapers. They also have historical use in basket making by Native Americans of the Southeast.

Longleaf pine trees produce the largest cones (far left) of all the southern pines, making them attractive for use in arts and crafts. Other pines shown from longleaf pine cone to the right are slash pine, loblolly pine, and shortleaf pine.

Longleaf Pine Cones. As with all pine trees, both the male and female parts are found on the same longleaf tree. The open pollination of longleaf pine cones usually begins in late winter (late February to mid-March). To the irritation of those with allergies or fastidious car habits, pollen begins to shed in the spring along with several other southern pines. Once fertilized, woody cones on the longleaf pine tree begin to grow and eventually mature by the autumn of the second year.

Cones open and seeds fall in October and November. Seeds are large and heavy and, despite having a wing to help with dispersal, generally do not fall far from the tree. Seeds are high in fats and are thus highly sought after by seed predators like mice, birds, squirrels, and ants. About every 7 years, longleaf pine trees have a mast period, producing cones in numbers too great to be eaten by seed predators. Partial crops of cones generally occur between these mast periods, but often all the cones are eaten.

The cones of longleaf pine are the largest of the southern pines and range in size from 5 to 12 inches long. Only animals like the fox squirrel are large enough to handle and open the longleaf pine cones to eat the seeds before they fall to the ground. Today, the large size of longleaf pine cones has made them attractive for use in arts and crafts.

Longleaf Pine Bark. On young longleaf pine, the bark is brownish gray and deeply creased. As the tree ages, the creases become shallower, the color assumes an orange-brown shade, and the exterior bark appears more paper-like. The thick bark of longleaf pine generally protects the trees from the frequent surface fires in the forest, but scaly bark tends to burn off. Generally, the soot of past fires can be viewed in the bark furrows. Occasionally, however, fires may find a weak spot and burn through the bark. From then on, the bleeding of resin from this wound renders future surface fires a great advantage, and over time a noticeable, indented fire scar may develop. A large fire scar may weaken, even kill, a longleaf pine tree.

It is common to find a variety of invertebrates such as centipedes, spiders, ants, and cockroaches seeking safety under the bark of longleaf pine trees. The Brown-headed Nuthatch (*Sitta pusilla*) and Red-cockaded Woodpecker are among the bird species that forage on the tree, peeling off scales of bark in search of those invertebrates. Some reptiles or amphibians like the barking treefrog (*Hyla gratiosa*) and pine woods treefrog (*Hyla femoralis*) take a more

The thick yet papery bark of the longleaf pine helps to protect the trees from frequent fires. This tree is about 50 years old.

Downed logs (known as coarse woody debris) and stumps play an important role in the longleaf pine forest. When a dead tree is still standing (called a snag or widow maker), it may be host to a variety of animal species. Southern pine (*Dendroctonus frontalis*) and black turpentine beetles (*Dendroctonus terebrans*), termites, and other insects feed on the decaying wood of the snag. In turn, woodpeckers and other critters feed on these beetles and their larvae. Woodpeckers often construct cavities in snags for roosting and nesting.

After abandonment, these cavities may become dens for southern flying squirrels (*Glaucomys volans*), Eastern Screech Owls (*Otus asio*), Eastern Bluebirds (*Sialia sialis*), Wood Ducks (*Aix sponsa*), and a host of other animals. When the tree falls, it will become home to dozens of other species. Oftentimes, reptiles such as snakes and lizards seek refuge in these downed trees and stumps. Eastern diamond-back rattlesnakes (*Crotalus adamanteus*) are known to use the downed logs as ambush points to catch small rodents.

Native American Burning Practices

Snags, downed logs, and stumps can provide important habitat for animals such as this corn snake (*Pantherophis guttattus*).

inactive approach to hunting food and simply wait on the bark for food to come to them.

Longleaf Pine Roots. Longleaf has a reputation of growing slowly during the seed and grass stages. However, while longleaf may look like it is not growing above ground, below ground it is very active. During the first few years following germination, longleaf pine seedlings quickly develop an impressive root system. Longleaf pine generally grows on dry, sandy soils (although they can be found on a wide array of soil types). In these dry, sandy

Longleaf pine trees typically grow a large tap root that helps reach deep soil moisture and also provides stability during strong winds.

soils, it is advantageous for the trees to quickly establish a root system that finds deep-lying soil moisture. Long roots also anchor the tree firmly in the ground.

Once longleaf pines start height growth, the large bundles of needles cause them to become somewhat top heavy. Without a stout tap root in place, these trees can easily fall over. Locals call this tumping over.

As the tree continues to mature, both the outward- and downward-growing roots continue to grow. In mature trees, roots spread outward an average of 35 feet from the trunk and even up to 75 feet. Longleaf differs from other pines in that the tap root is nearly as large in diameter as the tree trunk, tapering gradually to depths (on average) of 10–15 feet.

Because longleaf roots are high in some nutrients, various animals seek them out as food. Although relatively harmless to mature longleaf, pocket gophers excavate extensive caverns while feeding on the roots of longleaf pine and other plant species. The appetite of nonnative wild pigs for longleaf roots has resulted in countless areas of young trees being rooted up. When a longleaf pine tree dies, fires often consume the stump, creating a hole with the dimensions of the tap root. The burned or rotted-out root systems of these stumps create tunnels that are ideal for various cold-blooded creatures.

Stumps contain high resin content that made them attractive in turpentine production. This stumping, as it was called, was to the detriment of many reptiles, amphibians, and other animals that used these stumps and stump holes.

Longleaf Pine Forest Types

For thousands of years, various features of soil, elevation, and climate influenced fire behavior and forest growth. This, in turn, influenced the composition and structure of longleaf forests, including plant and animal species. These different forest types are often lumped into four groups: mountain, sandhill, rolling hills, and flatwoods and savannas.

Mountain Longleaf Pine. Longleaf pine dominates the south- and southwest-facing sides of the mountains and ridgelines up to about 2,000 feet in elevation in northern Alabama and northwestern Georgia. This habitat also includes an isolated example in Pine Mountain, Georgia, extending to Thomaston, Georgia, and in the Uwharrie Mountain Range in south-central North Carolina. Since much of the northern boundary of the range of longleaf pine is found in this habitat type, the forest changes from only longleaf pines to a mixture of longleaf pines and other southern pines and hardwoods. About 20% of the original longleaf pine ecosystem was covered by this mountain longleaf habitat type.

Sandhill Longleaf Pine. These habitats are characterized by ridges of loose sandy soils that begin in North Carolina and run through western Georgia between about 500 and 600 feet above sea level. A few isolated sand ridges exist on the Florida panhandle and peninsula and in southeastern Virginia. Longleaf pine sandhills are characterized as a forest of widely spaced pine trees with a stunted understory of deciduous (scrub) oaks and a sparse to continuous ground cover of bunchgrasses and herbs. Today, sandhill longleaf sites make up some of the largest acreages of remaining longleaf pine habitat (despite comprising roughly 10% of the original landscape).

Rolling Hills Longleaf Pine. These sites have the soils and sufficiently rolling topography to ensure good drainage. They are often very productive stands capable of producing excellent longleaf pine timber. In 1906, J. F. H. Clairborne described the longleaf pine forests in these rising and falling hills this way: "For twenty miles at a stretch in places you may ride through these ancient woods and see them as they have stood for countless years, untouched

Montane longleaf pine forests can be found on steep mountainsides growing upward of 2,000 feet in elevation, for example in North Georgia and Alabama.

Sandhill longleaf pine is often found with many deciduous, scrubby oaks. These deep sandy soils are typically thought of as the most suitable habitat for longleaf pine, in large part because it is one of the successful southern pines to grow in the sandhills, which largely haven't been converted to other pine species.

Longleaf pine forests in the rolling hills area of the Southeast are often quite scenic. Many of these areas have been converted to agriculture or production of other southern pines, principally loblolly pine.

by the hand of man and only scratched by the lightning or the flying tempest. This growth of giant pines is unbroken on the route." It is speculated that 30% of the original ecosystem was covered with this habitat type.

Flatwoods and Savanna Longleaf Pine. Characterized by level, poorly drained pine forests, flatwoods and savannas are often interspersed with frequent (and sometimes large) swampy patches or wet prairies. The primary distinguishing feature of savannas and flatwoods is the density of pines. As the name implies, a pine savanna (or pine prairie) is more akin to islands of widely scattered longleaf pines amid a sea of grass. Flatwoods sites can have an exceptionally dense cover of longleaf pine. However, depending on the

Longleaf pine and saw palmetto are frequently found growing together in longleaf pine flatwoods and low, often poorly drained areas. A large percentage of this habitat has been mechanically bedded and planted in other southern pine species, principally slash pine.

moisture in the soil, they can produce longleaf pine 40–90 feet tall. Flatwoods and savannas can have the highest diversity of groundcover grasses, herbs, and shrubs compared with the other longleaf habitats.

Since the soils are poorly drained and typically have low nutrient reserves, numerous orchids and carnivorous plants can be found here. This habitat type is found in seven of the nine states of the longleaf pine range and is often further described as the Atlantic coastal flatwoods or Gulf coastal flatwoods. Roughly 40% of the original landscape had this habitat type.

Sometimes called pine prairies, longleaf pine savannas often have a low density of pine trees but abundant grasses and forbs. Years of neglect through fire suppression have transformed many of these historically wet savannas to gallberry and titi thickets.

Fiery Disturbance

Following Fire

Fire is important in maintaining longleaf pine forests and the many plants and animals that make this forest their home. Historic fires were low in intensity. Frequent lightning storms and early humans were primary ignition sources for these fires.

Most plants and animals in the longleaf pine forest have developed behaviors or adaptations to survive in these areas with frequent fires. The gopher tortoise plays a key role in helping animals escape fire, as their burrows provide safe places to escape the flames. White-tailed deer (*Odocoileus virginianus*), Mourning Doves (*Zenaida macroura*), and Northern Bobwhite (*Colinus virginianus*) run or fly ahead of the flame front. The fox squirrel climbs a tree to the safety of the canopy while the fire passes underneath. Insects, like tiger swallowtails (*Papilio glaucus*), cloudless sulfur butterflies (*Phoebis sennae*), and American grasshoppers (*Schistocerca americana*), either fly ahead of the flame front or fly up to the safety of the tree crown. Some bird species take advantage of this insect smorgasbord.

The longleaf pine forest is well adapted to fire. The six small images at top show that ground-layer grasses and shrubs recover rapidly after a fire. Many of the resprouting plants are eaten by animals. The bottom image shows how fire is essential in the development of longleaf pine forests: lack of fire causes the forest to fill in with fire-sensitive shrubs and smaller hardwood trees that become abundant.

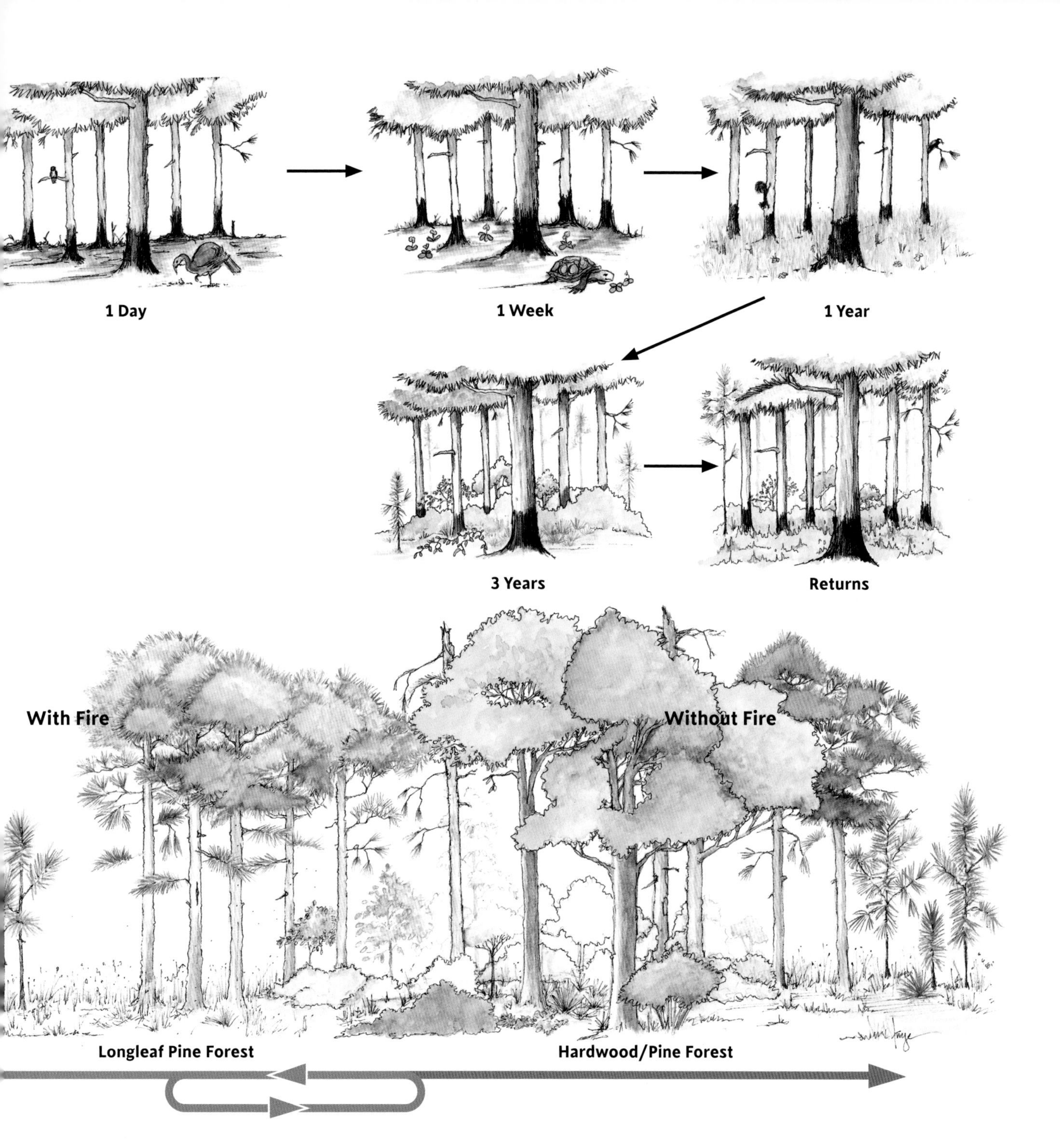
1 Day
1 Week
1 Year
3 Years
Returns
With Fire
Without Fire
Longleaf Pine Forest
Hardwood/Pine Forest

Canopy Disturbance

If you were to drive past a longleaf pine forest, you might think that all the trees are the same size. However, walking through the forest you can see tightly clustered patches of longleaf pine seedlings. Young longleaf pine seedlings actually take advantage of the forest openings created by the death of adult trees to regenerate. These disturbances occur somewhat regularly. Lightning kills one or two adult trees frequently in the longleaf ecosystem, while tornadoes and hurricanes may kill hundreds of trees less frequently. Scientists refer to these forest openings as forest gaps.

Lightning occurs frequently in the southern piney woods. Here, the wound caused by a lightning strike is visible on this longleaf pine tree. Such a strike often leads to the tree's demise. Lightning can also cause surface fires, mostly during the spring and summer.

CHAPTER 2

Natural and Human History of the Longleaf Pine Forest

Indigenous Stewardship

For thousands of years, the longleaf pine ecosystem was home for Indigenous cultures of the southeastern United States. As the Pleistocene era drew to a close about 12,000 years ago, warming weather marked the beginning of a new way of life and triggered cultural adaptations. With the extinction of large prey from overhunting and climate change, these Indigenous people conformed their hunting practices for the pursuit of smaller, more solitary game (like deer) that began to radiate out across the Southeast. With increased success in hunting and gathering, populations began to increase. The itinerant life of these hunter-gatherers began to change. Larger groups were becoming restricted to smaller territories. Bands of Native Americans began to focus their activities along specific water drainages.

The development of agriculture led to more reliable food stores, which in turn led to further increases in population. With the exception of the fertile red clay soils, it is doubtful that the deep sandy soils found in much of the upland coastal plain proved suitable for agriculture. Instead, the planting of maize, squash, and beans (called the three sisters) was generally relegated to the rich bottomlands near rivers and streams. By the time Spanish explorers arrived in the 16th century, many areas of the Southeast contained densely populated, sociopolitically complex chiefdoms. The longleaf pine forest remained a significant source of food, medicine, tools, housing materials, and clothing for most Indigenous cultures in the region.

Lightning-ignited fires played a role in creating desired supplies; however, because these fires were random where they burned (and thus unreliable), Indigenous cultures purposely burned areas to preserve many of the desired plants and animals in the forest around them. Over time, the frequent fires began to mold a forest of fire-tolerant longleaf pine and other plant species.

Burning by Seminole and other Indigenous peoples has a long history in longleaf pine forests.

European Arrival and Colonization

In 1539, Hernando de Soto led hundreds of Spanish and Portuguese soldiers, tradesmen, and priests into Florida. Traveling along well-established trails, de Soto and his men found their way from Indigenous town to Indigenous town in search of precious metals, often fighting along the way.

Soon after de Soto's expedition, other Europeans and enslaved Africans arrived in increasing numbers and the Indigenous populations plummeted in response. A primary reason for the decline is that these Indigenous people were susceptible to viruses and diseases that had been introduced from other continents. As the region's population of humans dropped, that of the white-tailed deer (*Odocoileus virginianus*) and other game animals (e.g., elk and bison) rose for a period. This spike in game animal numbers gave rise to a lucrative buckskin trade.

Colonists who settled the piney woods saw opportunities for raising cattle and other livestock. In the longleaf uplands, abundant bunchgrass furnished a cheap source of forage for the scrub cattle of these early southern cowboys (called cow catchers or cracker cowboys). When traveling through the wiregrass country of Georgia in the 1770s, American botanist William

Bartram noted "numerous herds of cattle" that were "peaceably browsing on the tender, sweet grass, or strolling through the cool, fragrant groves of the surrounding heights." By 1860, three-quarters of a century later, cattle and other southern livestock were worth almost half a billion dollars—more than twice the value of that year's cotton crop.

The open range herding practices of these pioneer hog drovers and cattle drivers of the piney woods were developed as a practical adaptation to the environment. Included in this ideology was a vehement opposition to enclosures. Typically, fences only enclosed a few acres of "cowpen" land on which subsistence crops were grown. Cows and hogs were simply turned out in the customary tradition of open-range herding. Range was often so unrestricted

Pineywoods Cattle

When European settlers arrived in the South, they found longleaf pine trees towering over a carpet of grasses as far as the eye could see. All of this grass provided a cheap source of food for the cattle. The longhorned cattle, brought by the early Spanish settlers to the New World from southern Spain, were allowed to roam freely and became semi-wild. Though open-range grazing has long been discontinued in the South, this endangered breed of cattle can occasionally be seen on farms scattered across the southern states.

They are lean and small, in the 600- to 1,000-pound range. Most have horns that range from small, Jersey-like curved horns to large, longhorn types. Their color varies among white, blue-sided, red-sided, red-speckled, black-speckled, strawberry-speckled, and solid colored.

These animals were often semiferal, developing a nature that was self-sufficient, tough, and hardy. Sometimes they are referred to as rakestraw because of their ability to survive the winter months by raking the longleaf pine straw back with their horns and muzzles to find grass protected from the frost. Some families recognized that older cows could be gentle and easy to handle once domesticated. Many of these cows were used to make good oxen to help log the longleaf pine forests years before modern tractors.

Transported from southern Spain, the Pineywoods cattle are known as a tough breed that were allowed to roam freely and became semi-wild in the longleaf pine forests.

that it was not unusual for animals to wander away beyond returning or finding.

The pastoral economy of this culture also encouraged the habitual use of fire, a tradition meshed with Indigenous fire management practices. Besides its use for maintaining palatable grasses for cows, woods burning was thought to be beneficial by killing snakes, ticks, chiggers, and fever germs. To keep wooden structures from being consumed by these frequent range fires, dirt yards were meticulously swept of debris. Woods burning was so common that the notable wildlife biologist Herbert Stoddard Sr. estimated that areas of Florida burned annually in many years.

Exploitation and Exhaustion

In the early decades of the 20th century, the longleaf pine region was responsible for producing 70% of the world's supply of naval stores—the collective name for products such as tar, pitch, spirits of turpentine, and rosin obtained from pine trees. A century earlier, North Carolina so dominated turpentine production that it earned the nickname Tarheel State for the black gummy tar that would accumulate on workers' bare feet. The highly resinous wood (often called fatwood) of the longleaf pine tree was responsible for sparking the naval stores industry throughout the Southeast—among the earliest industries in North America. The term "naval stores" was originally applied to the pitch and tar needed for waterproofing the wooden sailing vessels of the Royal British Navy in the 17th century. Around 1850, the production of gum turpentine peaked in North Carolina and began to spread south into Georgia as northerly forests were exhausted. However, it wasn't until after the Civil War that turpentine production began to rapidly increase.

Though longleaf timber had been cut in a few of the Atlantic states, after the Civil War longleaf forests were more intensely exploited for timber through the harvesting of stands bordering rivers and streams. Massive longleaf pines were felled with hand axes or crosscut saws, skidded to the water by oxen teams, assembled into rafts, and floated downriver to sawmills.

Gum from the pine tree was distilled into rosin and spirits of turpentine in what can be described as an oversized liquor still (not visible here). The collection and processing of pine gum was a year-round ordeal that often required a large labor force. Work was physically demanding and often began when the sun broke through the tree line and ended as darkness was cast across the sky.

Pitch and tar derived from the dry distillation of longleaf pine knots and logs. Though this process started immediately upon colonization, it continued into the early 20th century.

As the naval stores industry evolved, the French method of cupping live trees was later adopted starting in the early 1900s. The sticky gum removed from these longleaf pine trees was taken to a distillery to be made into products like rosin and spirits of turpentine.

With significant capital investments by northern investors, the efficiency of logging the virgin longleaf pine forest reached fever pitch in the early 1900s, and much of the original forest fell, such as in this scene from South Mississippi in the 1930s.

The exhaustion of forests elsewhere in the northern United States by the end of the 19th century saw northern investors purchasing large areas of southern lands covered with vast acreage of longleaf pine. With this investment came advances in technology to reduce costs and step up production. Tram lines for railroad logging penetrated the backcountry. Inefficient oxen teams were replaced with powerful, steam-powered skidders that could handle five or six logs at one time. Logging train engineers known as hoggers commanded flatcars loaded with longleaf pine logs down unsteady rail lines to colossal sawmills capable of cutting over 80,000 board feet per day.

Logging the exquisite longleaf pine forest that was once found on the north shore of Lake Pontchartrain. From 1908 to 1938 this wood fed the sawmill at Bogalusa, Louisiana, once the largest mill in the world.

The idea was simply to "cut out and get out," and by 1909 the peak of longleaf pine lumber production was reached. The extraordinary demand for every merchantable stick of timber, the pitiless system of taxing every board foot of standing trees, and the insatiable appetite of the piney woods rooter (feral hogs) assured that many areas were stripped clean of trees. Barren, cutover land stretched wearily away from rusting rail lines as far as the eye could see.

In this era, industry had little regard for forest conservation or sustainability.

During the Great Depression, Civilian Conservation Corps enrollees began to plant large areas of cutover land with pine seedlings, snuffing out the longleaf forest. The forestry community at the time saw longleaf pine as a slow-growing tree that was difficult to regenerate. So, in place of longleaf, loblolly or slash pines were planted in dense rows, amid the skeletons of the ancient longleaf pine forests. However, these seedlings were less tolerant of fire than the longleaf pine trees.

In order to protect this new investment of trees, a crusade spearheaded by the Southern Forestry Educational Project was initiated to preach that fire was a destructive agent in the landscape and needed to be snuffed out. Their mantra was by and large the following: So long as fire is kept out of the woods, the community of trees and plants and animals have a chance to take care of themselves. Later, the torch of fire suppression would be passed to Smokey Bear.

Heaving with books, pamphlets, and other outreach material, a fleet of trucks visited churches, schools, and supper clubs to preach the ills of fire in the South's piney woods in the late 1920s and early 1930s.

Utilizing a new technology called moving pictures, films were very effective in portraying fire in a novel, negative—and unfortunately misleading—light. This still is from the 1931 production of *Burning Bill.*

Longleaf pine in the rolling hills of southwestern Mississippi belonging to the Keystone Lumber and Improvement Company in 1901.

Longleaf pine forest in Rapides Parish, Louisiana, in 1940, belonging to the Bentley Lumber Company.

Old-growth longleaf pines outside of Holly Hill, South Carolina, 1954.

Longleaf pines in the mountains of Alabama (present-day Talladega National Forest), circa 1930s.

Flatwoods longleaf pines near Okeechobee, Florida, in 1925.

Sparse longleaf and slash pines outside Apalachicola, Florida, circa 1920s.

Revival

In a little over 150 years, the longleaf pine landscape went from a forest type dominating the southern landscape to a forest unrecognizable to many who live near it. The longleaf forest was in serious decline and at risk of completely disappearing unless actions for its revival were taken. Though warnings began in the early 1900s, they were slow to be heeded. By the late 20th century, concern had reached a frenzied pitch. Many groups and individuals joined forces to draw a line in the sand and actively work to promote and restore longleaf pine. Landowners, both public and private, are increasingly aware of the benefits provided by a healthy longleaf pine ecosystem and are investing in actions to restore and manage this diverse forest. Since about the year 2000, the acres planted in longleaf pine have been promising and have resulted in reversing this habitat's precipitous decline. Likewise, vigorous effort is going into managing the ecosystem in order to allow low-intensity fires, which can

Advances in seedling quality and planting techniques are aiding longleaf pine reforestation.

Early attempts at reforestation were hampered by poor seedling survival and the longleaf pine's slow growth. Other pine species were often planted instead, as in this circa–late 1930s photo.

help prevent high-intensity fires. Because of the ongoing efforts of partners across the Southeast, the future of longleaf pine forests looks bright.

Although remnants of this once-great longleaf pine forest still exist, they are often just a reminder of its former expanse and majesty, recognizable only to the informed observer. The following chapter introduces the next generation of outdoor enthusiasts to many of the unique animals and plants that our forebears may have known. Likewise, this information is intended to reacquaint those who have watched the slow demise of one of the world's most exquisite forests—the South's forest, the longleaf pine forest. Far from being complete, this guide highlights some of the more common animals and plants that can be seen in longleaf pine forests as well as some of the rarer and now-extinct species of this ecosystem.

Tools of modern reforestation of longleaf pine on lands where the species had been removed.

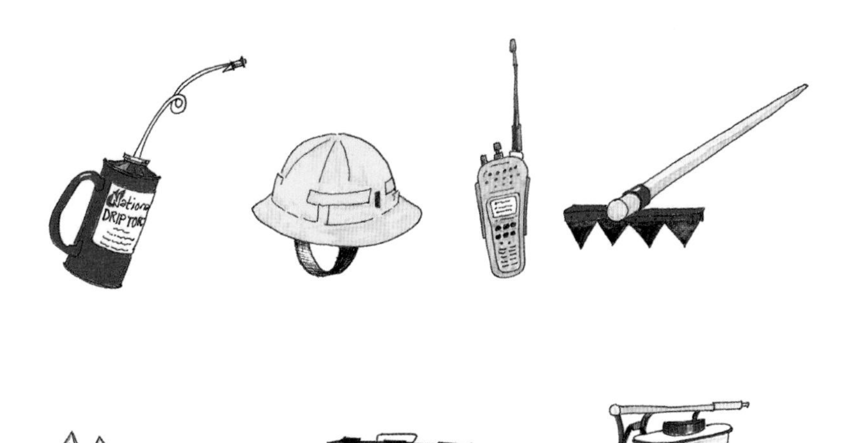

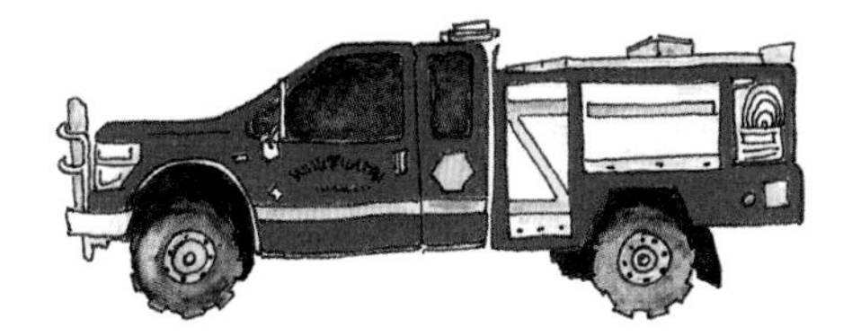

Tools of modern prescribed-burning practices in longleaf pine forests.

CHAPTER 3

Flora and Fauna of the Longleaf Pine Forest

Mammals

Shrews

Least Shrew (*Cryptotis parva*)
Southeastern Shrew (*Sorex longirostris*)
Southern Short-Tailed Shrew (*Blarina carolinensis*)

Often mistaken for field mice or moles, shrews are widespread across the longleaf pine range. Shrews can provide important food for other species, such as the Southeastern Kestrel. Of the shrews native to the Southeast, the three species typically encountered in longleaf pine forests include least shrew, southeastern shrew, and southern short-tailed shrew.

SIZE: The least shrew is the largest of the three species, at about 3½ inches long and weighing about ¾ ounce. The southern short-tailed shrew is next, at a little more than 3 inches in length and about ½ ounce. As the name implies, it comparatively has the shortest tail. Southeastern shrews are the smallest, at about 3 inches and ⅛ ounce. All three species have a pointed snout.

COLOR: Overall dark in color, the least shrews are grayish brown, southeastern shrews are reddish brown, and southern short-tailed shrews are typically slate gray. All have lighter colored underbellies.

BEHAVIOR: All these shrews have poor hearing and eyesight but movable whiskers around the nose and mouth that allow them to detect the vibrations of prey. They all feed primarily on invertebrates, such as crickets, grasshoppers, slugs, snails, centipedes, beetles, and spiders. Shrews are generally solitary, highly aggressive toward other shrews, and active at all hours of the

Least shrew.

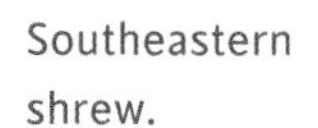

Southeastern shrew.

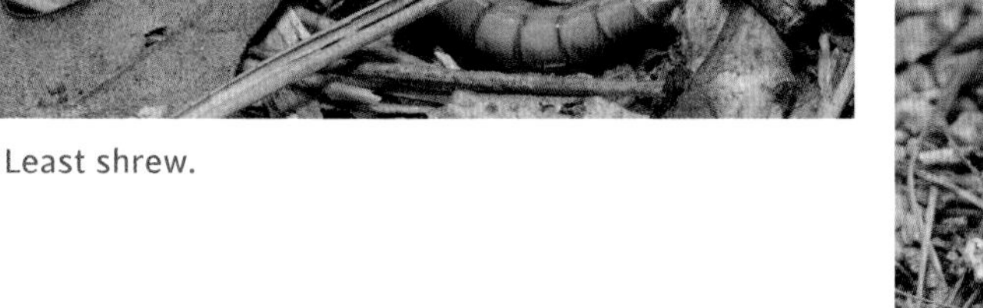

Southern short-tailed shrew.

day and night. They may burrow below rotting logs or within burrows made by moles and other animals. Thick overhead cover helps protect them from predators such as owls, snakes, and opossums. Strong-smelling musk glands can help deter predators in some cases.

ADDITIONAL NOTES: Though shrews were once accused of voraciously consuming longleaf pine seed and seedlings, research in the late 1960s proved otherwise. Instead, shrews focus primarily on insect food. Some shrews, including the southern short-tailed shrew, are venomous. These venomous shrews are able to eat vertebrates such as frogs and other rodents. Shrews do not have hollow fangs like vipers but instead possess glands above grooved teeth that can deliver the venom into small wounds created by a bite. Many shrews have extremely high metabolisms that make them voracious eaters. The southeastern shrew was formally called Bachman's shrew, in honor of the naturalist John Bachman, who described it in 1837.

Pine Vole

Microtus pinetorum

Pine voles spend most of their time in their tunnel systems, seldom venturing to the surface. Their elaborate tunnels and openings are the most conspicuous sign of their presence. These tunnels can be found in areas of dense organic matter where frequent fires have not occurred within pine forests. Tunnels can also be seen in thick grassy fields or pine and scrub oak with dense tree and shrub cover. Unlike moles (which also create belowground tunnels), pine voles have chisel-like front teeth that they sometimes use to chew the bark off the base of small longleaf pine seedlings, resulting in death of the tree. They are also known by the common names of pine mouse or woodland vole.

SIZE: This small mammal has shortened features: a short tail (about ½ inch long on many), very short ears, and a snub nose. Adults are about 4–5 inches long and weigh on average about 1 ounce. They have well-developed forelimbs and claws that assist them greatly in digging.

COLOR: Their short, velvety fur is reddish to chestnut in color.

BEHAVIOR: Their labyrinth of tunnels are often 1–2 inches below ground. Tunnels are often 1 inch in diameter. Within this tunnel system, pine voles eat, socialize, and breed. A single burrow can accommodate a colony of many

Pine vole.

adults and their young. Voles feed extensively on roots and stems of plants but can also eat fruits, seeds, and insects. Venturing aboveground exposes them to predation by hawks, owls, and some other larger mammals such as coyotes (*Canis latrans*), raccoons (*Procyon lotor*), and gray foxes (*Urocyon cinereoargenteus*). Oftentimes, pine snakes (*Pituophis melanoleucus*) will pursue and capture these animals in their tunnel systems. Predation is high enough that the life expectancy of the pine vole is often less than 1 year in the wild.

ADDITIONAL NOTES: Pine voles are known to cache large amounts of food underground in their tunnel systems.

Mice

Cotton Mouse (*Peromyscus gossypinus*)
Golden Mouse (*Ochrotomys nuttalli*)
Oldfield Mouse (*Peromyscus polionotus*)
Florida Mouse or Gopher Mouse (*Podomys floridanus*)

Mice are a diverse group of rodents that some people consider pests. However, most mouse species are essential in local ecosystems, playing a vital role in seed dispersal and as a food source for other animals. Four of the more common species of mice include cotton mouse, golden mouse, oldfield mouse, and Florida mouse. All but the last, which is found only in Florida, can be found throughout the range of longleaf pine.

SIZE: The cotton mouse body is on average 7 inches, with the tail adding another 3 inches. Golden mouse bodies are on average 6½ inches, with the tail being about 3 inches. The oldfield mouse body is on average 5 inches long, with the tail adding 2 inches. Florida mouse bodies are on average 8 inches long, with the tail adding 4 inches.

COLOR: Fur colors are as follows: dark brown for the cotton mouse; burnt orange for the golden mouse; dark brown for the oldfield mouse; grayish brown with orange flanks for the Florida mouse. All four species have a white underbelly.

BEHAVIOR: Both the golden mouse and oldfield mouse eat mostly grains and seeds. The Florida mouse and cotton mouse add berries, fruits, and arthropods to that diet.

ADDITIONAL NOTES: Though oldfield mice can be found in sandhill habitat, they often construct their own burrows. By contrast, the cotton mouse and Florida mouse may use burrows already created by gopher tortoises (*Gopherus polyphemus*). The golden mouse may build a nest off the ground using material such as Spanish moss.

Cotton mouse.

Oldfield mouse.

Golden mouse.

Florida mouse.

Bats

Northern Yellow Bat (*Lasiurus intermedius*)
Southeastern Myotis (*Myotis austroriparius*)
Eastern Red Bat (*Lasiurus borealis*)
Evening Bat (*Nycticeius humeralis*)
Tricolored Bat (*Perimyotis subflavus*)
Seminole Bat (*Lasiurus seminolus*)
Little Brown Bat (*Myotis lucifugus*)
Northern Long-Eared Bat (*Myotis septentrionalis*)

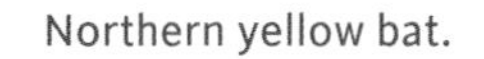

Northern yellow bat.

Eastern red bat.

Southeastern myotis.

As dusk begins to settle on the longleaf pine forest, it is not unusual to look up toward the darkening sky and see small creatures flying around. Typically, these are bats leaving their roosts in search of insects for food. Though several species of bats can be found in the longleaf pine forest type, eight species of bat—the northern yellow bat (aka Florida yellow bat), southeastern myotis, eastern red bat, evening bat, tricolored bat (formerly known as the eastern pipistrelle), Seminole bat, little brown bat, and northern long-eared bat—are notable inhabitants. Bats are major insect predators, consuming approximately one-third their body weight in insects per night. They commonly eat insects such as cicadas, damselflies, mosquitoes, beetles, and ants. A few species, such as Seminole and yellow bats, roost frequently in Spanish moss, while evening bats can be found roosting in longleaf pine snags. Some populations have been greatly affected by white-nose syndrome caused by the fungus *Pseudogymnoascus destructans*.

Northern Yellow Bat. This bat has a body length of 4½–5 inches, a wingspan of roughly 14–15 inches, and weighs ½–¾ ounce. Its fur is yellowish brown to yellowish orange.

Southeastern Myotis. The body length is 3–4 inches, its wingspan is about 9–11 inches, and its weight ¼–⅓ ounce.

Evening bat.

Tricolored bat.

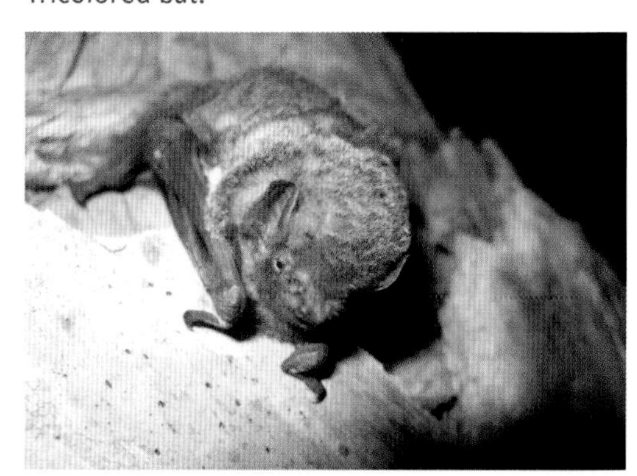

Seminole bat.

Little brown bat.

Its belly fur is typically white, which contrasts with brown fur on its back. Nose is described as pinkish.

Eastern Red Bat. The body length is 3½–4½ inches, its wingspan approximately 11–13 inches, and its weight ⅓–½ ounce. Its fur is usually brick red, though the female is duller in color.

Evening Bat. The body length is 3–4 inches, its wingspan is about 10–11½ inches, and its weight averages ⅓–½ ounce. It has dulled brown fur and black ears and wings.

Tricolored Bat. The body length is 2¾–3¾ inches, its wingspan is 8–10 inches, and its weight ¼–⅓ ounce. It has a black wing membrane that contrasts sharply with pink forearm bones. Its fur can be silver-yellow to dark orange.

Seminole Bat. The body length is about 4 ½ inches, its wingspan is 11–13 inches, and its weight approximately ⅓–½ ounce. The fur is a mahogany color with white tips that give it a frosted appearance. The shoulders and wrists have a distinguishing patch of white fur.

Little Brown Bat. The body length is 2½–4 inches, its wingspan is approximately 8½–11 inches, and its weight is ¼–½ ounce. Its fur is dark brown and glossy on its back. The underbelly is grayish.

Northern Long-Eared Bat. The body length is 3–3¾ inches, its wingspan is 9–10 inches, and its weight averages ¼–⅓ ounce. Their fur color is medium to dark brown on the back and lighter brown on the underside.

Northern long-eared bat.

Southeastern Pocket Gopher

Geomys pinetis

Though sometimes mistaken for fire ant mounds, small mounds of sand thrown up at the forest surface indicate the presence of pocket gophers. It is known by other common names such as sandy mounders or salamanders (not to be confused with the group of amphibians). Living most of its life belowground, this animal itself is rarely seen. These burrowing animals once had several distinct isolated populations throughout the longleaf pine forests, including one just outside of Tuscaloosa, Alabama, earning this area the nickname of Salamander Hills.

SIZE: These creatures measure about 10 inches in total length, weighing on average 8½ ounces, and have small eyes, ears, and tail. Males are slightly larger than females.

Southeastern pocket gopher in a rare aboveground photo.

COLOR: Short fur is brownish gray.

BEHAVIOR: Pocket gophers use their powerful front legs and feet with long claws for digging an extensive belowground burrow system. Feeding on a diet of fleshy roots that grow into its tunnels, the pocket gopher is able to close its lips behind the protruding front teeth, allowing it to chew underground without getting dirt in its mouth. Poor eyesight and hearing do little to protect the pocket gopher from its enemy, the pine snake. Instead, these animals use dirt to plug the tunnels to slow the movement of this snake predator.

ADDITIONAL NOTES: Some people consider pocket gophers a keystone species since their tunnels are home to many arthropods. Some insects, like certain dung beetles, require these tunnels and would cease to exist without the pocket gopher. The Baird's pocket gopher (*Geomys breviceps*) is found in the western range of longleaf pine.

Hispid Cotton Rat

Sigmodon hispidus

This native rodent species is widespread across the longleaf pine range and does very well in areas dominated by a thick grassy understory.

SIZE: Adult average total length is about 12 inches and weighs approximately 6 ounces. Its ears are relatively short, extending a very short distance above fur. The relatively short tail is thinly haired.

COLOR: Long, coarse fur on the back is a mixture of brown, tan, and black while the underside is grayish or white.

BEHAVIOR: The diet is primarily green vegetation, including high-protein legumes and seeds of many grass species. They will occasionally eat insects and the eggs of ground-nesting birds such as quail and turkey. They have been documented to use gopher tortoise burrows. However, they are often in heavily vegetated areas, where overhead herbaceous cover is sufficient enough to protect them from predators. They will create well-defined runways under the grass that usually radiate out from a nest of ball-shaped woven grasses. They have many offspring but only live on average about 6 months in the wild. They are eaten by hawks and owls, bobcats, coyotes, and snakes (corn snakes are a major predator).

Hispid cotton rat.

ADDITIONAL NOTES: Cotton rats have been shown to feed on pine seeds and seedlings. Grass-stage longleaf may be killed if partially or completely girdled near the root collar.

Southern Flying Squirrel

Glaucomys volans

This nocturnal animal is rarely seen by humans. The species is arboreal, selecting habitat in trees and avoiding the ground. In the daytime, they often roost in cavities such as those created by woodpeckers. The cavities in dead trees are often made by Red-headed (*Melanerpes erythrocephalus*) and Red-bellied Woodpeckers (*Melanerpes carolinus*) and in live trees by Red-cockaded Woodpeckers (*Picoides borealis*). Many animals, including birds and snakes, may compete for this cavity space.

SIZE: Body length is 8–10 inches and weight 2–4 ounces.

COLOR: The back is grayish brown; the belly is white and has a densely furred (and flattened) tail.

BEHAVIOR: Flying squirrels do not actually fly. Instead, folds of loose skin between the wrists and ankles allow them to glide up to 250 feet—or perform

Southern flying squirrel.

a controlled fall, using the tail for guidance. Though the smallest of tree squirrels, it is the most carnivorous, with its diet including some birds and nestlings as well as eggs, carrion, and invertebrates. Being nocturnal allows them to escape many predators like hawks.

ADDITIONAL NOTES: These animals can be problematic for Red-cockaded Woodpeckers and are known to occupy and (in some cases) prey on existing nests.

Fox Squirrel or Southern Fox Squirrel

Sciurus niger

Fox squirrels are the largest tree squirrels in North America. This large body size is an adaptation to help survive in the open landscape of certain eastern forests, including longleaf pine ecosystems. Size advantages might include defense against being carried off by predators and retaining higher energy stores.

Fox squirrel.

SIZE: Fox squirrels generally measure 26–41 inches in overall length from tip of nose to end of tail. The weight is usually 1–2 pounds.

COLOR: Several subspecies of fox squirrel are found in longleaf pine's range. Color can vary between subspecies and even specific populations within subspecies. Some are black with silver or white markings around the feet, nose, and the tips of the ears and tail. Others are silver-gray, with black markings. A more reddish variety of fox squirrel is typically found in the mountain longleaf pine habitat.

BEHAVIOR: Fox squirrels eat a variety of foods, including acorns, berries, tender buds of plants, arthropods, fungi, and the seeds of the longleaf pine itself. The large longleaf seeds are a favorite because they are nutritious and their high fat content is a good source of energy. The feeding squirrels often leave behind piles of cone scales and the skinny core of the cone. The fox squirrel's size helps it manipulate the large and heavy cones of the longleaf pine.

ADDITIONAL NOTES: Fire is what separates good fox squirrel habitat from gray squirrel (*Sciurus carolinensis*) habitat. Without frequent fires to suppress oak trees and prevent other shrubs from taking over, the forest type will change and gray squirrels will outcompete their cousins, the fox squirrel.

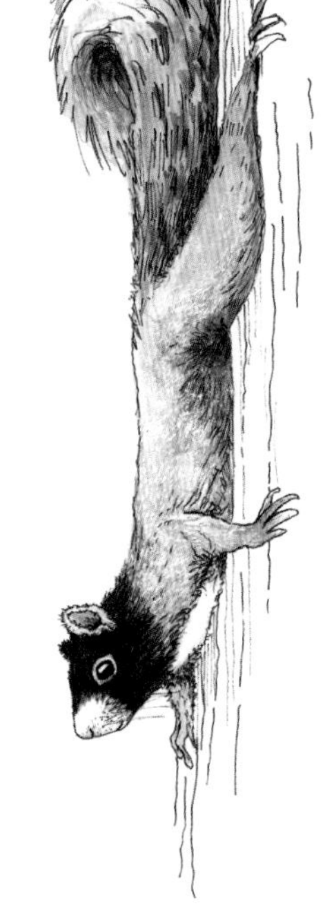

Fox squirrels are often found scampering through healthy longleaf forests. The coloration can vary from light gray to brown to all black.

Striped skunk.

Striped Skunk

Mephitis mephitis

Often smelled more than they are seen, skunks are well known to most people. A solitary animal, the striped skunk can be found denning in longleaf pine forests using burrows created by armadillos, gopher tortoises, or foxes. They are also known locally as polecats.

SIZE: The small head and short legs gives this skunk a weasel-like appearance. The plumed tail, about 1 foot long, is distinct. The total body length averages 25 inches.

COLOR: The body is black with two white stripes down the back. The stripes themselves can vary from the entire back being white to no stripe being present at all. The forehead has a distinctive white stripe.

BEHAVIOR: Striped skunks have musk glands that can produce a foul-smelling stream up to 15 feet; this can be smelled up to 1 mile away. Great horned owls (*Bubo virginianus*) are unfazed by this musk and are their primary natural predators. Vehicle collisions also account for a large number of deaths. Skunks forage in longleaf pine forest for a variety of foods, including insects, reptiles, amphibians, roots, bird eggs, and fruits such as muscadine grapes (*Vitis rotundifolia*) and persimmons (*Diospyros virginiana*).

ADDITIONAL NOTES: A cousin of the striped skunk is the eastern spotted skunk (*Spilogale putorius*). Though black and white and of similar size, spotted skunks do not commonly occupy the pine-dominated forests, which frequently burn.

Virginia Opossum

Didelphis virginiana

The opossum is adaptable to a wide variety of habitats, and it's not uncommon to see one lumbering through a longleaf pine forest. Virginia opossums occupy a similar niche as raccoons and compete with that species for food. However, opossums may select smaller prey and are a major predator of ticks. Most locals refer to this species as just opossum or possum.

SIZE: The opossum is about the size of a large house cat. Male opossums can weigh on average 8 pounds, while smaller females are around 4½ pounds. The average length from nose to end of tail is about 2½ feet.

COLOR: Opossums are generally gray with flecks of black fur distributed throughout. The muzzle is white and the nose pink. Ears are black with white tips. On rare occasions, opossums can be all white.

BEHAVIOR: The opossum is North America's only marsupial. The average number of young it produces (about 8) enter the mother's pouch after birth and remain there for 50–85 days. As the young wean and become large enough, they will hitch a ride on their mother's back. Agile climbers, opossums may find dens in trees. They can also be found in burrows such as those constructed by gopher tortoises. Opossums are known for scavenging dead animals but can also be known to eat a variety of food—from eggs, arthropods, birds, and amphibians to berries and nuts. In turn, opossums may fall prey to large owls, foxes, coyotes, and bobcats. When startled, opossums may pretend to be dead (play possum): the tongue hangs out, the lips curl back to show many of its fifty teeth, it salivates, and it emits a foul order from a gland.

Virginia opossum.

ADDITIONAL NOTES: Opossums were a vital food source in the U.S. South. Historically, sweet potatoes and possum (possum and taters) or possum crisp and brown gravy were a common component of southern cuisine. Benjamin Harrison had two pet Virginia opossums in the White House, Mr. Reciprocity and Mr. Protection, which might originally have been obtained as food for a White House Thanksgiving dinner.

Gray Fox

Urocyon cinereoargenteus

These animals are a common resident of longleaf pine forests and are seen typically in early morning or evening hours. A subspecies of gray fox was once found along the Gulf Coast yet has probably bred with foxes imported from outside the region.

SIZE: Adults are 32–60 inches from nose to end of tail, with an average weight of 8 pounds. Males are on average larger than females. The kits typically reach full size within 5–6 months.

COLOR: Fur is gray on the upper body, with the lower parts whitish. Legs, neck, and back are rusty yellow. The tip of the tail is black, which distinguishes it from gray phases of the red fox (*Vulpes vulpes*).

BEHAVIOR: These foxes are omnivores, feeding on a variety of food sources in longleaf pine forests, from the oldfield mice to persimmon fruit that have fallen from the tree. They have been known to den in the root wads of tipped-up longleaf pine trees and the abandoned burrows of gopher tortoises. These sneaky creatures are more likely to hide from predators rather than run.

ADDITIONAL NOTES: Foxes are members of the canine, or dog, family. Sometimes the gray fox is mistaken for the red fox, mainly due to similar size and its reddish-brown fur. Gray foxes are one of the few canines that are adept at climbing trees.

Gray fox.

Red fox.

Nine-banded Armadillo

Dasypus novemcinctus

Though they have long occupied the piney woods of East Texas, armadillos are a relatively recent introduction to the remainder of longleaf pine forests, having migrated on their own from west of the Mississippi River. Having few natural predators and the ability to give birth to four identical young has allowed the armadillo to spread quickly throughout the South. With poor eyesight and hearing, this is one animal in the longleaf pine forest that is easy to approach for viewing. They also go by the names of nine-banded armadillo and long-nosed armadillo.

SIZE: Their average length is 15–20 inches and average weight is 9 pounds.

COLOR: A grayish red outer shell of overlapping scaly armor covers the back, sides, head, tail, and outside surfaces of the legs. The underbelly and inner surfaces of the legs have no armor protection but are covered by tough skin and a layer of coarse hair.

Nine-banded armadillo.

BEHAVIOR: These animals are often found feeding at dusk. The exact impact of armadillos on other native species is still under study, but it is not uncommon to see armadillos using gopher tortoise burrows or rooting up the nest of a reptile to eat its eggs. Armadillos have even been documented rooting up planted longleaf pine seedlings. When they are not foraging, armadillos amble along, stopping occasionally to rise onto their hind legs to sniff the air for trouble.

ADDITIONAL NOTES: It is a myth that this species of armadillo can roll up into a ball. Instead, to startle predators, nine-banded armadillos will often jump straight up in the air, sometimes 3–4 feet off the ground. If this doesn't work, they will quickly flee or bury themselves in a way that exposes only the hard upper shell.

Raccoon

Procyon lotor

The raccoon is well known in the Southeast. Although it is more common in other habitats, it is not unusual to see one in longleaf pine forests. An intelligent animal with an appetite for eggs, the raccoon can be problematic for some species in decline, such as Northern Bobwhite (*Colinus virginianus*) and gopher tortoises. Raccoons will typically be seen at night, roaming in search of food.

SIZE: A stocky, medium-sized mammal averaging 35 inches in overall length, the raccoon averages 15 pounds. The distinctive bushy tail has several black rings on it. The face narrows to a pointed nose.

COLOR: A bandit's black mask outlines the eyes that, in turn, are outlined in white. The coat is a grayish color with hints of brown and black.

BEHAVIOR: The long, thin, hairless, flexible front fingers and paws are distinguishing features on raccoons. This feature allows them a great deal of dexterity to grasp and hold objects. Their claws are not retractable. Raccoons sometimes wash their food before eating it. If no water is available, the raccoon can rub off any debris found on their food. They prefer food that is easier for them to catch, like arthropods, eggs, fish, and crayfish. They also eat a good deal of plant material, such as acorns and berries.

Raccoon.

ADDITIONAL NOTES: Like many small mammals, raccoons were hunted extensively for their meat and pelts until the mid-20th century, and are still hunted and trapped for recreation and as part of predator control efforts.

Bobcat

Lynx rufus

These medium-sized felines have a short, or bob, tail, that gives them their common name. Especially secretive, they are not easily seen by most observers. Catlike in appearance, these compact predators in longleaf pine forests keep animals such as rodents from becoming overpopulated. Locals may refer to them as wildcats.

SIZE: The adult bobcat is roughly 2–4½ feet from nose to stubby tail. Males are slightly larger than females, averaging 30 pounds for males and 23 pounds for females.

Bobcat.

COLOR: The body, forelegs, and tail are tan to grayish brown, with black streaks on the body and dark bars on the forelegs and tail. They are also spotted with black patterning, which helps camouflage them. The lips, chin, and underside are generally off-white. The ears are black-tipped and pointed. The nose is often pink.

BEHAVIOR: Due to the higher abundance of prey in longleaf pine forests than in other habitats, the area used by bobcats is often less here. The area varies for males (toms) and females (queens) and between seasons, with males generally requiring twice as much territory as females, or up to 1,500 acres here (that's nearly 1,500 football fields). Bobcats have been known to feed heavily on cotton rats, mice, rabbits, young white-tailed deer (*Odocoileus virginianus*), and birds.

ADDITIONAL NOTES: A group of bobcats is called a clowder or clutter.

White-tailed Deer

Odocoileus virginianus

These animals are perhaps the best-known inhabitants of southern forests. However, by the early 1900s white-tailed deer were exceptionally rare, having been hunted to near extinction. The abundant numbers today are a testament to excellent restocking and management efforts.

SIZE: The female (doe) is about 36 inches tall at the shoulder, whereas the males (bucks) of similar age are slightly taller. Within the longleaf pine's range, a healthy adult doe weight ranges from less than 90 to up to 140 pounds, while bucks weigh 140–250 pounds. Bucks grow antlers each year, often reaching their largest size around 5–6 years. At birth, most fawns weigh 4–8 pounds.

COLOR: Summer coats are short, thin, and reddish brown. This coat is replaced by a longer grayish-brown coat with a short, wooly undercoat. Their belly, chest, throat, and chin are white throughout the year. The coats of newborn fawns are reddish brown with several hundred small white spots. This spotted coat helps conceal them in the grassy understory of longleaf pine forests.

BEHAVIOR: These are extremely adaptable animals that can thrive in a variety of habitats. These animals are browsers and can often be seen eating the

White-tailed deer buck.

White-tailed deer doe.

tender new growth of plants following fires. It is not uncommon to see the bark of a young longleaf pine sapling scrubbed off by the buck in the wintertime during rutting (breeding) season.

ADDITIONAL NOTES: The trade of cured deer hides (buckskin) was one of the first industries in the South. Native Americans traded buckskin with Europeans for hatchets, trade beads, fabric, and other items. This trade expanded, driving many deer populations to near extinction. For example, in 1775 exports of deer hides out of Pensacola, Florida, totaled 30,000 deer. Fire was frequently used to help round up these speedy animals.

Black bear.

Black Bear

Ursus americanus

This species does not necessarily require longleaf pine forests. However, they prefer large expanses of uninterrupted woodlands, which may include longleaf pine forests, especially in national forests and wildlife refuges. There are over 16 subspecies, three of which are found in the Southeast and include the eastern, Florida, and Louisiana black bears. Of those, the eastern and Florida black bears have made tremendous comebacks over the past few decades.

SIZE: The size of the bear varies by sex, time of year, age of the animal, health of the animal, and the subspecies. They are large, bulky animals. Males weigh on average 300 pounds and females 150 pounds. Adults stand on average 24–36 inches at the shoulder and can be 60–72 inches long.

COLOR: Shiny black fur covers most of the body. Some may have a white blaze on the chest. The muzzle has short brown fur.

BEHAVIOR: Described as reclusive animals, they're more apt to show up on trail cameras than be seen walking about. The thicker the vegetation and the more remote the area, the more these animals seem to flourish. But they can range over large areas. These animals have a diverse diet of plants, animals, and insects. They slow down in the winter and den for short periods, but unlike the subspecies farther north are not forced into hibernation.

ADDITIONAL NOTES: The black bear had spiritual significance to many Indigenous cultures. Many Creek Indian towns maintained a *No'kose em ekana*, or "beloved bear ground," for these animals.

Coyote

Canis latrans

Coyotes (also known as song dogs) have made a migration eastward and are now found widely distributed across much of the Southeast.

SIZE: Coyotes weigh 20–50 pounds. They stand on average 1½ feet tall at the shoulder. They have a narrow snout with a long tail.

COLOR: Coyotes are light gray or brown. On rare occasions, solid-black coyotes can be seen.

Coyote.

BEHAVIOR: Coyotes do well in close proximity to people. It is not unusual to see coyotes in suburban and even urban environments. The coyote diet reflects its adaptability and often includes rodents, rabbits, birds, eggs, many kinds of fruit, domestic poultry, pets, and human food waste. Coyotes serve a function of preying on smaller animals and/or sick and weak animals. They have also been documented to help disseminate seeds such as persimmon and pawpaw (*Asimina triloba*). Coyotes have been known to eat the eggs of gopher tortoises and Northern Bobwhite, to the detriment of those populations.

ADDITIONAL NOTES: Coyotes may form family groups or packs. However, coyotes are more apt to hunt and travel alone, giving the impression that they are solitary animals.

Scary Things in the Woods

The piney woods of the South have always been a place of folkloric haints or mystical creatures. Some of these creatures originated from the imagination of logging camps and others have just been around, passed through the generations.

The splinter cat (*Felynx arbordiffisus*) is a creature described by many early loggers in the region. This elusive animal is rarely seen and is harmless to humans. Often, what a person sees is the aftermath of powerful, nocturnal feline feeding. A stand of longleaf pine with several trees snapped off several feet above the ground may be the work of a hungry splinter cat rather than the wind. While searching for food along the trunks of trees, the splinter cat has a propensity to snap off trees while hunting around for insects or small mammals such as raccoons. Have you seen a stand of longleaf pine with several trees snapped off several feet up the tree? It may not be a tornado or hurricane, but instead the work of a hungry splinter cat.

The hoop snake (*Serpenscirculousus caudavenenifer*) is found throughout North America but finds ideal habitat in the piney woods. Particularly aggressive, the snake will grasp its tail in its mouth and roll after its prey like a hoop or wheel. With the open nature of longleaf pine forests, these snakes are said to be able to build up great speed. The tip of the hoop snake tail contains a stinger that the snake will straighten at the last minute to strike its prey. If you see this snake rolling after you in the woods, know that they do not turn well, so you can protect yourself by hiding behind a tree. Note, however, the poison is strong enough that if the hoop snake's tail strikes the longleaf pine, it will likely die.

Splinter cat.

Bardin Booger is a large primate that is particularly fond of the longleaf pine forests around Central Florida. This large, upright creature is thought to be related to what the Seminole called *Esti Cap Caki*, meaning "large hairy man" or a southern variety of Bigfoot. Coweta County, Georgia, has the Belt Road Booger, which is perhaps a hairy cousin of the Bardin Booger and another primate known as the devil monkey. Hairy primates in the deep woods of the South have long been intertwined in the lore.

Where Have All the Bison Gone, When Will They Ever Herd?

The abundance of plants and animals ebbs and flows with the changes in longleaf pine forests. Many of the animals listed in this guide are sensitive to small changes such as those seen after a fire or after the canopy has been disturbed by a hurricane or logging. For example, if the forest goes too long between fires, certain birds will quickly relocate elsewhere. Larger, landscape-level changes such as forest fragmentation can permanently impact some animals, causing them to be unable to reinhabit these forests. Mammals such as American bison (*Bison bison*), red wolf (*Canis rufus*), and Carolina yellow dog (*Canis lupus familiaris*) all used to make their homes in longleaf pine forests.

Early European explorers and settlers frequently mentioned bison in the southeastern United States. Following closely behind fire, bison took advantage of the

American bison.

Carolina yellow dog.

Red wolf.

lush grassy savannas and prairies. In turn, their shaggy coats probably helped transport seeds that helped further shape this ecosystem. The American bison replaced an ancient bison (*Bison antiquus*) that died out 12,000 years ago, possibly due to overhunting by early hunter-gatherers.

Rediscovered in the 1970s in the longleaf pine woods of South Carolina, characteristics of the Carolina yellow dog (also called the American dingo) led researchers to believe that this animal descends from the semidomesticated dogs kept by the original Native Americans of the coastal longleaf pine forests. All remaining wild dogs were captured in the 1980s as part of a breeding program, so it is unlikely that this animal would be seen now in longleaf pine forests. However, their role in this forest remains unknown.

For thousands of years, red wolves are believed to have roamed over much of the longleaf pine landscape. Persecuted and hunted, their numbers declined. By the 1980s, only about 14 red wolves remained in the wild. Those animals were captured and put into captive breeding programs to keep them from going extinct. The ever-adaptable coyote moved in and filled their niche. Coyotes (*Canis latrans*) and red wolves are known to hybridize (mate with each other and create offspring). Research continues to determine where red wolves fall on the family tree of North American canids and whether or not they represent a distinct species.

Birds

Northern Bobwhite

Colinus virginianus

This bird is known locally as bobwhite quail or simply quail. The bobwhite name comes from the call of the male Northern Bobwhite, a common sound in healthy longleaf pine forests. The Northern Bobwhite depend on fire in the longleaf pine forest to maintain abundant ground cover and expose bare ground so they can scratch around for food.

SIZE: Average adults have a body 10 inches long and a wingspan of 15 inches. They can weigh up to 9 ounces.

COLOR: Northern Bobwhites are brown, buff, white, black, and gray, with males having a white throat and a brow stripe bordered by black.

Male Northern Bobwhite Quail.

Hunting Bobwhite Quail.

BEHAVIOR: Northern Bobwhite generally feed heavily just before dark. The majority of an adult bird's diet consists of vegetable matter. Of this vegetable matter, the seeds of fire-dependent native legumes are highly desired (especially during the winter season, when grass seeds have spoiled). The reverse is true of birds 2 weeks old or younger. To help young birds grow quickly, their diet consists mainly of protein-rich arthropods that they hunt for in "bugging areas." Fire helps maintain bugging areas where the chicks are able to move around and feed.

ADDITIONAL NOTES: At the turn of the 20th century and during the Great Depression, large tracts of land in South Georgia, Central Alabama, and areas of the Carolinas were purchased by wealthy business leaders who wanted to establish areas to hunt quail. Today these quail hunting preserves represent some of the best remaining examples of longleaf pine forests.

Ground Dove

Columbina passerina

This bird is well camouflaged among the dull-colored grasses and dusty ground where it is found. Frequently, the presence of these birds is unknown until one explodes into flight, creating a whirring sound. Ground Doves, like other members of the dove family, have weak feet and are poor scratchers for food in heavy litter or vegetation. Fire maintains shallow litter layers and open sandy stretches, giving birds easy access to a variety of native grasses and plant seed. Also known as sand dove, moaning dove, or tobacco dove.

SIZE: These doves are relatively small, with a body length of 6–7 inches, wingspan of 11 inches, and weight averaging 1–1½ ounces.

COLOR: Ground Doves have dark scaling on the chest and neck but otherwise are a dull brown all over. Males have a pinkish coloration on their necks.

BEHAVIOR: These birds spend a tremendous amount of time on the ground feeding and nesting. If not for their inconspicuous coloration, they probably would be easy prey for the many predators found in longleaf pine forests, including opossums, raccoons, and hawks.

ADDITIONAL NOTES: It is speculated that ground doves must eat a few thousand seeds of grass per day in order to meet their calorie requirements.

Ground Dove.

Energy-packed native pea seeds found frequently in burned longleaf pine forests can help make up for the lower nutritional value of grass seeds. The Mourning Dove (*Zenaida macroura*) is also common in the longleaf pine forests. Though both dove species feed on seeds on the ground, you're more apt to see the Mourning Dove perched on the branch of an old pine tree.

Yellow-billed Cuckoo

Coccyzus americanus

This slender, dove-sized bird with a long tail is elusive, often remaining motionless while sitting on its perch. Its presence is more often heard than seen. This cuckoo has a very distinctive knocking call that it uses day or night. Some have described the knocking as a guttural cuck-cuck-cuck repeated several times. Since tent caterpillars on oak and cherry trees are a primary food source, a good place to listen for Yellow-billed Cuckoos is near an infestation of those insects.

SIZE: Both sexes are of similar size, with a body length about 10–12 inches and weight about 2–2⅓ ounces.

COLOR: It has a grayish brown back, head, and upper tail. The wings are reddish brown. Breasts and underparts are white. The underparts of the tail have black and white bars. The bills are yellow and curved downward.

BEHAVIOR: This bird builds poorly constructed nests from small sticks and debris. In some circumstances, it lays its eggs in other species' nests, especially the Black-billed Cuckoo (*Coccyzus erythropthalmus*). When this nest is threatened by a predator, one of the pair will fly to a nearby perch and pump its tail up and down while opening its wings.

ADDITIONAL NOTES: It is said that Yellow-billed Cuckoos tend to call more right before a rain, thus earning them the nickname of Rain Crow or Storm Crow.

Yellow-billed Cuckoo.

Common Nighthawk

Chordeiles minor

This medium-sized bird is exceptionally well camouflaged. For this reason, it is often seen (and heard) in the air. They can be seen flying looping patterns, swooping, and darting, especially at dusk or dawn. During the daytime they sit or perch mostly motionless on either the ground or a tree branch. It is also known as a bull bat or mosquito hawk.

SIZE: Both sexes are of similar size, with a total length of 8½–9½ inches and a weight of 2¼–3½ ounces.

COLOR: Both sexes are highly camouflaged with a general drab color. They have dark barring on the sides and underside and white patches on the wings. In flight, a V-shaped white patch is seen on the throat.

BEHAVIOR: The feeding behavior of this bird is fascinating. Although the bird itself has a small bill, it has a large gape when the mouth is fully open. It flies with the mouth open and catches prey, mostly insects, on the wing. Nesting is also unusual: nests are no more than small indentations on the ground with a few sticks and other debris hastily scattered around the eggs, but often Common Nighthawks nest on bare ground. The eggs are exceptionally camouflaged, and only the most observant individual is fortunate

Common Nighthawk.

Common Nighthawk nest.

enough to see this bird on the ground. Parents of the nest will occasionally perform elaborate displays to lure potential predators away from the eggs, often pretending to have a broken wing.

ADDITIONAL NOTES: This bird has earned the name Bull Bat in large part because of its flight. Some people suggest it flies erratically like a bat. Moreover, males in courtship display will go into a steep dive during flight. The rushing wind through the wings makes a distinctive whirling or booming at the bottom of the dive. These birds are also springtime migrants to the longleaf pine forests. Scientists have found some nighthawks that breed in Florida spending the winter in Argentina, 4,000 miles away.

Vultures

Black Vulture (*Coragyps atratus*)
Turkey Vulture (*Cathartes aura*)

The two Vulture species common in the Southeast are sometimes found together. Both are large birds of prey and are erroneously called buzzards. They are both scavengers and are often seen around roadkill or dead animals (earning them an additional name of Carrion Crows).

SIZE: Black Vulture males and females are the same size, with a total length of about 1¾–2½ feet. Wingspan is about 4¾–5½ feet. Turkey Vultures are generally larger, with a body length of about 2–2½ feet and a wingspan up to 6 feet.

COLOR: Both species are large and predominately black. Turkey Vultures have red skin on their featherless heads, while Black Vultures have dark gray-black heads. In flight, Turkey Vultures have a mostly uniform coloration, but the wings may look light on the rear half as light filters through the feathers. This contrasts with the Black Vulture's darker overall feathers, but they have a distinctive lighter band toward the outer portion of the wing.

BEHAVIOR: The Turkey Vulture's incredibly strong sense of smell allows it to smell food (dead animals) from up to 1 mile away. Black Vultures, on the other hand, rely more on eyesight and look for other vultures circling

Black Vulture.

Turkey Vulture.

over the carrion to find their next meal. Black Vultures hunt cooperatively and will kill and eat calves, white-tailed deer, and other much larger animals. The birds are easy to distinguish by their flight patterns. Black Vultures vigorously flap their wings and soar for short distances before flapping again. Turkey Vultures appear more graceful in flight and are able to sail long distances. Black Vultures often congregate in single species groups, while Turkey Vultures tend to be less social.

ADDITIONAL NOTES: Both vulture species have very interesting (some say gross) adaptations to their lifestyles. Their digestive juices kill bacteria, allowing them to eat rotten meat. It is also speculated that they both have bald heads to avoid infections related to consuming rotten meat. Finally, they've been known to defecate on their legs. Some say this is to help keep them cool, and others say it acts as a sanitizer.

Red-shouldered Hawk

Buteo lineatus

This medium-sized hawk is regularly seen or heard in longleaf pine forests. People may also call this species striped hawk or chicken hawk. Unlike its cousin, the Red-tailed Hawk (*Buteo jamaicensis*), which circles high above the forest, the Red-shouldered Hawk hunts from a perch and thus is often seen on tree limbs, fence posts, or power-line poles. Though the call is unique, Blue Jays (*Cyanocitta cristata*) sometimes mimic it, which can make auditory identification a challenge.

SIZE: Males are 17–23 inches total length and weigh on average 1¼ pounds. Females are larger, at 19–24 inches and averaging 1½ pounds.

COLOR: Red-shouldered Hawks have a distinctive reddish brown on their shoulders. Adults have a dark brown head except in Florida, where the head is often pale gray. The back is brown, while the chest area is reddish. The belly is white with dark brown streaks. The tail and wings are barred black and white. A white crescent on the wings is a distinctive identification characteristic when in flight.

BEHAVIOR: These birds aren't particular in what they eat. They've been documented feeding on small rodents, large insects, fish, frogs, and toads. Snakes also make up some component of their diet. They have great eyesight, and their hunting style is helped by the unobstructed views in longleaf pine forests.

ADDITIONAL NOTES: The chicken hawk name arises because of its appetite for smaller birds, including chicken and quail. For this reason, these hawks (and others with the same appetite) were once shot on sight. In some areas of South Georgia, they are simply called quail-eating "rascals."

Red-shouldered Hawk.

Eastern Screech Owl

Megascops asio

This is the smallest owl species encountered in longleaf pine forests. A master of camouflage, it is typically heard but not seen. If seen in the daytime, it is often positioned in a tree cavity facing outward. These birds have two distinctive tufts on the top of their heads that may look like ears but are, in fact, feathers. Since they are nearly silent in flight, the sound often heard is their call during the twilight hours. This owl makes a high-pitched whinny, or screech, as the name implies.

SIZE: Males are slightly smaller but on average these owls are 6½–10 inches in total length and weigh approximately 4⅓–8⅔ ounces.

COLOR: These owls can have two general colors known as phases: mostly gray or mostly reddish brown (rufous). Their bodies also have complex patterns of bands and spots to give the bird excellent camouflage. They have bright

Eastern Screech Owl with red morph coloration.

yellow eyes. Some experts have suggested the reddish brown owl is more common in longleaf pine forests, perhaps due to the color of the pine trees.

BEHAVIOR: These birds are strictly nocturnal. They feed predominantly on mice, shrews, and voles though they have been known to eat insects on occasion. The prey is swallowed whole and later pellets of indigestible bones and fur are regurgitated.

ADDITIONAL NOTES: Unable to excavate their own cavities in trees in which to roost or nest, they instead rely on the abandoned cavities left by woodpeckers or holes opened or enlarged by rot or fungus. The female does not prepare an organized nest and will instead lay her eggs on whatever debris is at the bottom of her nesting cavity.

Woodpeckers

Downy Woodpecker (*Dryobates pubescens*)
Hairy Woodpecker (*Dryobates villosus*)
Red-bellied Woodpecker (*Melanerpes carolinus*)
Red-headed Woodpecker (*Melanerpes erythrocephalus*)
Northern Flicker (*Colaptes auratus*)
Pileated Woodpecker (*Dryocopus pileatus*)

Red-cockaded Woodpeckers have become the most iconic birds of longleaf pine ecosystems, but several other woodpeckers may be more commonly encountered throughout longleaf pine's range. These species include the Downy Woodpecker, Hairy Woodpecker, Red-bellied Woodpecker, Red-headed Woodpecker, Northern Flicker, and Pileated Woodpecker. Arthropods are the primary diet of most woodpeckers, but some species may also consume plant materials. These plant materials, such as nuts and other seeds, are especially important in the winter. Woodpeckers don't sing like most birds but use calls and drumming to indicate their territories. These other woodpeckers build nests in dead or decaying trees. Pileated Woodpeckers may sometimes excavate cavities in living trees where previous wounds contain softened wood. Other bird species may use the nest created by these woodpeckers.

Downy Woodpecker.

Hairy Woodpecker.

Red-bellied Woodpecker.

Downy Woodpecker. This small woodpecker is 5½–6½ inches long and weighs ¾–1 ounce. It has a short bill and is largely black and white, with white spots on the wings. It has a black-and-white-striped head. Males have a red nape.

Hairy Woodpecker. This medium-sized woodpecker is 9 inches long and weighs 1½–3½ ounces. It is often described as a slightly larger Downy Woodpecker with a longer beak.

Red-bellied Woodpecker. This medium-sized woodpecker is 9–11 inches long and weighs 2–3 ounces. It has a black-and-white zebra back and white underbelly. Females have a gray crown, while the males are red.

Red-headed Woodpecker. This medium-sized woodpecker is 7½–10 inches in length and weighs 2–3½ ounces. It is tricolored with a red head and neck, black back and tail, and white underparts and wing ends.

Red-headed Woodpecker.

Northern Flicker.

Pileated Woodpecker.

Northern Flicker. This medium-sized woodpecker is 11–14 inches in length and weighs 3–6 ounces. Its overall color is light brown (buff). It has a black crescent on its chest. The head is gray, and there is a red crescent on the nape. The undertail and underside of its wings are yellow. Males have a black mustache near the bill. Other names include Yellowhammer and Golden-winged Woodpecker.

Pileated Woodpecker. This large and conspicuous woodpecker can be up to 20 inches in total length. Bodies are black, with white on wings and face. A prominent red crest occurs on top of their heads.

ADDITIONAL NOTES: Another woodpecker, the Yellow-bellied Sapsucker (*Sphyrapicus varius*), can be seen in the winter in these forests and looks similar to Hairy, Downy, and Red-cockaded Woodpeckers. The Sapsuckers' horizontal rows of sap-filled wells can be seen on older longleaf pine trees.

Red-cockaded Woodpecker

Picoides borealis

Red-cockaded Woodpeckers are unique among eastern woodpeckers because they excavate cavities exclusively in mature, living longleaf pines (and a few other pine species). Other species excavate cavities only in dead trees and limbs. The pines that Red-cockaded Woodpeckers select are infected with red-heart disease, a fungus (*Phellinus pini*) that causes the dense inner wood of the pine to decay and become soft enough to construct a cavity. Longleaf pine trees generally have insufficient fungus-softened heartwood until 80–120 years old. Even then, it often takes a Red-cockaded Woodpecker 1–3 years to excavate through the living portion of the tree to reach the softened interior. In many cases, the birds that start excavating a cavity will never use it. Cavities are often completed by successive generations of related birds. Once the cavity is completed, however, it may serve as a roost or nest cavity

Red-cockaded Woodpecker.

for more than a decade. Red-cockaded Woodpecker cavities also are used by dozens of other species, making this woodpecker important for the overall diversity of cavity-nesting birds. In some old-growth forests, cavities excavated by Red-cockaded Woodpeckers occur at a density of one cavity per 5 acres. Generally, these birds will excavate groups of cavity trees in an area called a cluster. In 1970, the species was listed as endangered. Recent progress has raised hope of recovery of this bird.

The Red-cockaded Woodpecker is the only woodpecker species in North America that creates cavities in living pine trees. These cavities provide safe nesting and rearing habitat for its young.

SIZE: Their total length is on average 8½ inches and the wingspan is 14–15 inches. The average adult weight is 1½ ounces. Males are slightly larger than females.

COLOR: This bird has a black-and-white-barred back, noticeable white cheek patches, and a black cap and neck. Males have an inconspicuous red spot of feathers behind each eye (called cockades).

BEHAVIOR: Male helper birds from the previous nesting season help incubate the eggs and raise the young of the next generation. While this may seem altruistic, the young males are waiting for the dominant males in neighboring territories to die so they can compete to take over the territory. Frequent fires maintain an open and parklike forest preferred by the woodpeckers. These open conditions are also very favorable for a variety of invertebrate prey for Red-cockaded Woodpeckers.

ADDITIONAL NOTES: Gray rat snakes (*Pantherophis spiloides*) are agile tree climbers and are the primary predators of these woodpeckers. As a defensive behavior, the woodpecker chips small holes (called resin wells) into the bark of the cavity tree. These resin wells cause gum to ooze down the face of the tree. When the scales of the snake come into contact with this gum, the snake generally retreats back down the tree.

American Kestrel

Falco sparverius

This falcon is found throughout the Southeast. It also goes by the names of sparrow hawk and tillie hawk. Found in open pine woodlands and prairies, it prefers the drier longleaf pine forest types such as sandhills. A nonmigratory subspecies called the Southeastern American Kestrel (*Falco sparverius paulus*) is found year-round from Louisiana to Florida.

SIZE: This is the smallest falcon species in North America. The total length is roughly 9–12 inches and the wingspan is 20–24 inches. Females are slightly larger than males, weighing on average 3½ ounces (with males at 3¼ ounces).

COLOR: Males have blue-gray wings, while females have less impressive rusty-brown coloration. Both sexes have a black-and-white facial pattern and a black band at the end of a rusty-brown tail. Both sexes have buffy white, or off-white, undersides with black flecking; some biologists suggest the Southeastern Kestrel has less flecking than the American Kestrel.

BEHAVIOR: These birds can frequently be seen perching on trees or power lines, with their tail bobbing rapidly up and down. They hunt over areas with short grass and sparse ground cover typical of those seen in longleaf pine sandhills. They feed primarily on lizards and arthropods such as grasshoppers, beetles, and spiders. They have occasionally been known to feed on small rodents, amphibians, and birds. Pairs typically mate for life.

ADDITIONAL NOTES: Kestrels are called secondary-cavity nesters. This means they do not excavate their own cavities but instead rely on woodpeckers and other species to create a cavity. A decline in dead, large, pine trees has resulted in a lack of suitable cavities for kestrels. As a result, they are a species of conservation concern throughout their range.

American Kestrel.

Great Crested Flycatcher.

Great Crested Flycatcher

Myiarchus crinitus

This bird species travels by night to Central America and South Florida in order to spend the winter. However, in April–July (breeding season), it returns and can often be heard and seen high up in the treetops in the longleaf pine forests. It is rarely seen on the ground. Adult males and females are indistinguishable in appearance.

SIZE: This medium-sized bird is approximately 6¾–8½ inches in total length. They weigh about 1–1½ ounces.

COLOR: Its breast is a distinctive sulfur to lemon-yellow color. Its bushy crest is gray brown. It has a long, rusty-brown tail, and its sides are browner.

BEHAVIOR: Its diet largely consists of invertebrates such as beetles, spiders, moths, and butterflies. However, this bird species can supplement its diet with fruits such as grapes and blackberries. It hunts for insects from a perch, swooping down after its prey.

ADDITIONAL NOTES: Like several other birds in this field guide, this bird species is a secondary-cavity nester, making its nest within a cavity abandoned by woodpeckers. Like the Tufted Titmouse (*Baeolophus bicolor*) and Blue Grosbeak (*Passerina caerulea*), it is known to weave snakeskin into its constructed nest within the cavity.

Eastern Kingbird

Tyrannus tyrannus

This long-distance migrant winters in locations as far away as South America. However, it is typically seen or heard in the breeding season (spring/summer) in the longleaf pine forests due in large part to its personality. This fierce little bird is often described as aggressive, belligerent, and a tyrant.

SIZE: Male and female birds are similar in appearance. Considered a medium-sized bird, it is 7½–9 inches long and weighs 1¾ to about 2 ounces.

COLOR: The back, nape, head, and tail are dark gray. The belly, throat, and tail tip are white.

BEHAVIOR: It can often be seen perched on a tree limb while either feeding or defending its territory. From this perch, it flies off and catch beetles, grasshoppers, wasps, and other insects midflight. Sometimes it may hover in place before taking the insect. These birds are also seen regularly driving away predators or other birds.

ADDITIONAL NOTES: The Eastern Kingbird was named in 1758 by Carl Linnaeus, known as the "father of modern taxonomy" for his formalizing the binomial nomenclature system. Its color and size are similar to the Eastern Wood Pewee (*Contopus virens*) and the Eastern Phoebe (*Sayornis phoebe*).

Eastern Kingbird.

Loggerhead Shrike
Lanius ludovicianus

These birds are similar in size and appearance to the Northern Mockingbird (*Mimus polyglottos*) seen throughout the South. What the Loggerhead Shrike lacks in size, it makes up for in spirit. This fierce predator thrives in open grasslands with scattered trees, earning it the nickname butcherbird. Years of neglect and fire suppression continue to cause a decline in proper habitat for Loggerhead Shrikes. It is relatively rare to see this bird in today's forests.

SIZE: It is the size of a mockingbird, with a total length of 8–9 inches, wingspan of 11–13 inches, and weight of 1–1¾ ounces.

COLOR: This gray bird has black wings, tail, and face mask. The wing patches, throat, and outer edges of the tail are white.

BEHAVIOR: These birds are predators. A large part of their diet consists of grasshoppers and other large arthropods, but they have also been known to

Loggerhead Shrike.

Impaled prey of a Loggerhead Shrike.

kill lizards, snakes, small rodents, and other birds. They typically catch and impale their prey on sharp thorns, prickles, spines, or barbed wire. Though they require habitat with grassy understories like those found in longleaf pine forests, they prefer to nest in shrubby thickets such as those small pockets that fire may not reach. These birds can be found perching on a nearby tree or post and will dive down on unsuspecting prey. They may defend large territories.

ADDITIONAL NOTES: The hooked bill is more than just a fashion statement: they use the bill to break the back of prey such as lizards and snakes.

Blue Jay

Cyanocitta cristata

These are commonly found throughout most of the eastern and central United States. When flying through open areas of pine forests, they tend to stay quiet. However, once perched they can be raucous, producing a variety of musical notes. They have an uncanny ability to mimic the call of a Red-shouldered Hawk.

Blue Jay.

SIZE: Adult jays are 9–12 inches in length and weigh on average 2½–3½ ounces.

COLOR: They are lavender blue with white accents. The wings, back, tail, and face are often midblue. The cheeks are white, and the chest can be white with a darker gray. They have a black nape around the neck and black bands on the tail.

BEHAVIOR: In longleaf pine forests, these birds have a connection to many species of oak trees. Blue Jays may be responsible for a significant portion of oak dispersal, but more research is needed to understand this relationship. They have a fairly broad diet that includes various arthropods (grasshoppers, beetles, spiders, etc.), frogs, and even the eggs of other birds.

ADDITIONAL NOTES: Distinguishing males from females is very difficult. Their coloration is the same (monomorphic). Though males are slightly larger than females, size is not a good means to distinguish the sex. Their behavior differs, though: males can typically be seen making food offerings to females. During courtship, several males will compete for the attention of one female.

Carolina Chickadee.

Carolina Chickadee

Poecile carolinensis

This small, adaptable bird is found in a variety of habitats, including urban areas. You are just as likely to see one at your bird feeder as in the longleaf pine forest. These birds can often be seen hopping along tree branches, gleaning insects from the bark of the longleaf, or hanging upside down from a cone. They spend the majority of their time in trees or elevated structures but occasionally may forage on the ground.

SIZE: Adults are approximately 4½ inches in length and are the smallest of the North American chickadees. They weigh 3¼–4¼ ounces.

COLOR: The black-capped head and white cheeks are a distinctive feature. They have a short, black bill and small yet plump body. The wings, tail, and back are gray; the belly is white.

BEHAVIOR: The diet can include a variety of insects, seeds, and fruits. These chickadees do not create their own nests but instead occupy cavities created by nuthatches or woodpeckers.

ADDITIONAL NOTES: The famous John James Audubon first noted this species while in coastal South Carolina, calling it the Carolina Titmouse in his *Ornithological Biography*: "My drawing of the Carolina Titmouse was made not far from New Orleans late in 1820. I have named it so, partly because it occurs in Carolina, and partly because I was desirous of manifesting my gratitude towards the citizens of that State, who by their hospitality and polite attention have so much contributed to my comfort and happiness, whenever it has been my good fortune to be among them."

Tufted Titmouse

Baeolophus bicolor

This small songbird is often found flitting from branch to branch in longleaf pine trees. Sometimes it can be seen hanging upside down from cones like an acrobat.

SIZE: With a large head and thick neck, these birds look larger than other birds it may flock with in the forest. The total length is 6–6½ inches and weight is approximately ¾ ounce.

Tufted Titmouse.

COLOR: This bird is a gray, soft silver color with a distinctive rust-colored patch on its flanks. Just above the bill is a black patch. Its pointed crest is a distinctive feature.

BEHAVIOR: These little birds have big personalities and are well known for their fierce scolding of other titmice, other birds, or predators. They can also be found in flocks with other birds such as chickadees or nuthatches.

ADDITIONAL NOTES: They will take up residence in nest cavities left behind by Pileated Woodpeckers, flickers, and other woodpeckers. In this cavity they build a cup-shaped nest from grass, moss, wood fibers, or fur from animals such as raccoons, opossums, and fox squirrels. They've been documented plucking fur from living animals.

Brown-headed Nuthatch

Sitta pusilla

This is one of the few birds found exclusively in the United States. Its range follows that of pine forests from Virginia to Texas. This species thrives in open pine forests that are burned with regular frequency, such as longleaf pine ecosystems. Like many of the birds that require these habitat conditions, they are declining in numbers.

SIZE: This small songbird has a total length of about 4½ inches and wingspan of 6–7 inches, and weighs about ⅓–⅔ ounce. Males and females are similar in size and appearance.

COLOR: Both sexes have a short tail with a sharp, black, nail-like beak. The head has a brown crown with a distinctive white patch on the back of the neck. The lower part of the face is white, as is the throat and underparts. The remainder of the body is blue gray.

BEHAVIOR: The Brown-headed Nuthatch can be found feeding on mature longleaf pine trees, hopping along tree trunks and branches, hanging upside down, or climbing headfirst down trees. It uses its sharp beak to excavate seeds and catch insects. It uses existing cavities in snags for nesting and roosting. These birds will sometimes be seen using a chip of bark held in the beak to try to glean insects from the side of a longleaf pine tree.

ADDITIONAL NOTES: The Red-breasted Nuthatch (*Sitta canadensis*) is a

Brown-headed Nuthatch.

similar-looking member of the same family and can be seen or heard in the winter months in longleaf pine forests. The White-breasted Nuthatch (*Sitta carolinensis*) is also a resident in longleaf pine forests but also ranges widely through other forest types.

Carolina Wren

Thryothorus ludovicianus

This is a common wren species found in the eastern United States. In the Southeast, it's not uncommon to see and hear this aggressive little bird, as they are keen to defend their territory. It is the male bird that can be heard singing. Once it pairs with a female, it is usually mated for life. They are also known as great Carolina wren, mocking wren, or Florida wren.

SIZE: It has a total length of 5–5½ inches. It weighs about ⅔–¾ ounce.

Carolina Wren.

COLOR: Both males and females are reddish brown on the back, nape, and head. The breast area is buffy orange. The chin and throat are white. They have a long white stripe on the eyebrow area.

BEHAVIOR: These birds are often found feeding on the ground but can be seen moving up tree trunks in a way similar to nuthatch species. They hold their tail upright. The curved shape of the bill of the Carolina Wren is useful to turn over decaying vegetation. It can also use its bill to chisel and hammer apart large bugs. Its diet consists primarily of insects (ants, bees, beetles, moths, crickets, etc.), but spiders and snails may be gleaned off the sides of trees on occasion. Nesting usually takes place in abandoned cavities or natural holes in old snags.

ADDITIONAL NOTES: South Carolina claims the Carolina Wren, a year-round resident, as its official state bird.

Gray Catbird

Dumetella carolinensis

This medium-sized bird selects shrubby areas where it can forage for food undetected. They are rarely seen making long flights over open areas. Unlike Northern Mockingbirds, and more like Carolina Wrens, Gray Catbirds tend to hold their tails upright.

SIZE: Both sexes are slightly smaller than a Northern Mockingbird at 8½–9¼ inches in total length. Average weight is 1–2 ounces.

COLOR: Both males and females share the same color pattern, which is primarily slate gray. They have a reddish chestnut area under the tail, a black head cap, and a black tail.

BEHAVIOR: Though these birds eat a variety of insects, they also have a fondness for fruits. It is not uncommon to see catbirds gorging on

Gray Catbird.

the violet-colored berries of the American beautyberry (*Callicarpa americana*).

ADDITIONAL NOTES: Depending on location within the longleaf pine range, Gray Catbirds may be found year-round or migrating to warmer southern climates in the late fall and winter. Of their various calls, a distinctive catlike mewing indicates their presence. They can mimic other birds with an impressive repertoire of learned songs. If you get too close to an occupied nest, the male catbird will often perch, give the mewing call, raise its wings, fluff up its feathers, spread its tail, and open its bill skyward.

Northern Mockingbird
Mimus polyglottos

This common, medium-sized bird has a long tail, short, rounded wings, and a slight down-curved bill. Despite their prevalence today, in the 1800s their numbers had declined significantly because of collection as pets. Today, aside from abundance in suburban areas, they can be found in more open areas or

Northern Mockingbird.

edge areas of the longleaf pine forest. They can often be seen and heard sitting atop a perch and are known to aggressively defend their territory.

SIZE: Males and females are approximately the same size. The average total length is 8½–10 inches and the weight is about 1½–2 ounces.

COLOR: Males and females are indistinct, with an overall slate gray color. The breast and belly are pale gray. The tips of the wings are black. They have a white patch on their wings. The outside of the long tail feathers is also white.

BEHAVIOR: When not perched, these birds can be seen aggressively running or hopping after food. Their diet is heavy on insects, including beetles, grasshoppers, and ants, and infrequently includes small reptiles. Some fruits, such as serviceberry, can attract multiple individuals even if they're in disputed territory.

ADDITIONAL NOTES: Males are known for their singing repertoire. Some birds have been known to learn up to 200 different songs. Each song is repeated three times before they progress to the next mimic. This trait helps identify them in the field from other mimics (e.g., Gray Catbird). Their singing caught the attention of John Audubon, who stated that "there is probably no bird in the world that possesses all the musical qualifications of this king of song, who has derived all from Nature's self."

Eastern Bluebird

Sialia sialis

A few decades ago, researchers noticed Eastern Bluebirds had drastically declined, prompting efforts to recover their populations. Today, thanks to those recovery efforts, they are not that unusual in longleaf pine woods or open fields and farmlands.

SIZE: The average adult has a total length of about 6¾–8¼ inches, a wingspan of 10–13 inches, and weight of about 1 ounce.

COLOR: Males have a brilliant blue on top of the head and back, with a reddish-brown throat and breast. Adult females are spared such bright colors since they spend most of their time on the nest. Females have lighter blue wings and tail, a brownish throat and breast, and a gray head and back.

Male (right) and female (left) Eastern Bluebirds.

BEHAVIOR: The majority of this bird's diet consists of arthropods, including grasshoppers, beetles, and spiders. Fruits such as berries from flowering dogwood (*Benthamidia florida*), pokeweed (*Phytolacca americana*), and Virginia creeper (*Parthenocissus quinquefolia*) can supplement their diet in the winter months when insects become scarce. Rat snakes prey heavily on the eggs of Eastern Bluebirds, while the introduced European Starlings (*Sturnus vulgaris*) can compete aggressively for nesting cavities.

ADDITIONAL NOTES: During the breeding seasons, three other blue-colored birds can be seen in longleaf pine forests: the male Blue Grosbeak, the male and female Blue Jay, and the male Indigo Bunting (*Passerina cyanea*). Although all are bright blue, they each have distinct calls, foraging strategies, and habitat preferences.

Bachman's Sparrow

Peucaea aestivalis

Sometimes called a pineywoods sparrow, this bird is more often heard than it is seen. Its presence is a sign of a healthy longleaf pine forest. If fire is kept out and the mid-story shrubs begin to encroach, this bird will leave the area. Like so many other grassland birds in this region, Bachman's Sparrows are becoming more difficult to find.

SIZE: This small sparrow averages about ¾ ounce and is 5½–6 inches in total length. Wingspan is 2–2½ inches.

COLOR: The breast is buff colored with a whitish belly. The upperparts and head are rufous brown, with gray and black streaking on the neck and back. The face is gray, with a rufous-brown eye stripe.

BEHAVIOR: These sparrows require areas with a grass-dominated ground layer and open mid-story such as in well-managed longleaf pine forests. Breeding females will create a domed nest on the ground near a clump of bunched grass or a small shrub where she will lay 3–5 eggs. Adults feed heavily on seeds but may also rarely consume invertebrates, especially when nesting.

Bachman's Sparrow.

ADDITIONAL NOTES: Sparrows are often collectively referred to as LBJs (little brown jobs) by some birdwatchers due to the difficulty of species identification. Bachman's Sparrows are unique as far as sparrows go, but other visually similar species in longleaf pine forests include Henslow's Sparrow (*Ammodramus henslowii*), Swamp Sparrow (*Melospiza georgiana*), Song Sparrow (*Melospiza melodia*), Chipping Sparrow (*Spizella passerina*), Field Sparrow (*Spizella pusilla*), and White-throated Sparrow (*Zonotrichia albicollis*).

Eastern Towhee

Pipilo erythrophthalmus

These birds are common in longleaf pine forests, especially where the forest edge meets another habitat, an area called the ecotone. If the habitat is pure grassland with scattered longleaf as far as the eye can see, then you probably won't see an Eastern Towhee. However, if there are large, unburned patches or other areas with thick shrubs, then Eastern Towhees seem to thrive. In these shrub thickets, however, this bird is typically heard rather than seen. It is also known as the rufous-sided towhee or joree.

Male Eastern Towhee.

Female Eastern Towhee.

SIZE: It has a body length of 7–8 inches, wingspan of 9–11 inches, and weight averaging 1–2 ounces.

COLOR: Male coloration is more distinctive than the female, with a dirty-black head, upper body, and tail (these areas are brown in the female). The sides are rusty (rufous) colored and the eyes are distinctively red.

BEHAVIOR: These birds are not very particular about what they eat, consuming seeds, fruits, and insects of many varieties. When conditions are quiet, it is not uncommon to hear towhees scratching around in the leaf litter for food. Their thick, triangular beak allows them to crack open seeds that other birds in this forest type cannot. The female typically builds her nest on the ground, where it is frequently susceptible to nest parasites, such as Brown-headed Cowbirds (*Molothrus ater*), which lay their eggs in place of the towhee's egg. Ground nests may be destroyed by fire but, like many birds, Eastern Towhees can renest during the same season.

ADDITIONAL NOTES: These birds use both feet to scratch the ground at the same time. It looks like a backward hop.

Common Yellowthroat

Geothlypis trichas

This high-energy, small warbler can be seen darting around on low marsh grasses and shrub thickets. They are quick to scold intruders. They are more common in wet pine savannas and marsh areas.

SIZE: Both males and females are the same size, at approximately 4½–5 inches in total length and weighing about ⅓ ounce.

COLOR: Females are a drab olive with bright yellow under the throat and tail. Males have a black face mask highlighted with white and overall brighter coloration.

BEHAVIOR: This bird's diet is almost exclusively made up of insects that may be plucked from the ground or vegetation (gleaning) and caught in flight (flycatching). They can be year-round residents in some areas (mostly coastal) and migratory in other areas of the Southeast.

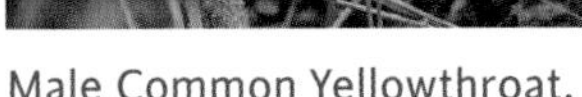

Male Common Yellowthroat.

Female Common Yellowthroat.

ADDITIONAL NOTES: Their nests are on or near the ground, which requires creative defenses against predators. To deter predators, Common Yellowthroats are known to land nearby and sneak to their nest.

Pine Warbler

Setophaga pinus

These are common in longleaf pine forests. However, as they tend to stay near the tops of the trees, they are typically heard rather than seen, without the assistance of binoculars. The best chance of seeing them without binoculars is in an area consisting mostly of longleaf pine saplings. They are small and active and thus difficult to photograph. These birds can be a year-round resident throughout most of the longleaf pine range.

SIZE: Males and females tend to be of similar sizes, with a total length 5–5½ inches, wingspan 9 inches, and weight ⅓–½ ounce.

COLOR: Males have a bright yellow throat and chest, while females (and young) birds are generally duller in color. The bellies are white with white wing bars.

Male Pine Warbler.

Female Pine Warbler.

BEHAVIOR: These birds can frequently be seen on branches in the upper canopy of longleaf pine trees. Here they can be seen eating large quantities of seeds and insects. They will put seeds into bark crevices and hammer them open with their long, pointed bill. These warblers will make their nest by gathering broomsedge, pine needles, and twigs to form a cup. Plant fibers and spider webs can be used to further bind the nest together. The nests are placed high on the longleaf pine tree near the end of a branch.

ADDITIONAL NOTES: The Pine Warbler and Common Yellowthroat are the only two members of the wood warbler family considered year-round residents of longleaf pine forests. However, at different times of the year, breeding season or winter, a number of different warblers belonging to the same family can be found in longleaf pine forests, including the Hooded Warbler (*Setophaga citrina*), Orange-crowned Warbler (*Leiothlypis celata*), Palm Warbler (*Setophaga palmarum*), Prairie Warbler (*Setophaga discolor*), Yellow-rumped Warbler (*Setophaga coronata*), and Yellow-throated Warbler (*Setophaga dominica*). Pine Warblers (though smaller) are easy to confuse with female Orchard Orioles (*Icterus spurius*).

Frequent-Flier Miles

Many of the birds described in this guide are known as residents, meaning they are found year-round in longleaf pine habitat. However, these forests can also be home to migratory bird species that may only be found here seasonally. Migrations are often triggered by impending food scarcity associated with a change of seasons. Some bird species such as Henslow's Sparrow (*Centronyx henslowii*), Yellow-bellied Sapsucker (*Sphyrapicus varius*), and Palm Warbler (*Setophaga palmarum*) migrate from the northern latitudes and spend the winter in the longleaf pine forest. Other birds (known as neotropical migrants) successfully migrate over the Gulf of Mexico to return to longleaf pine forests in the spring in order to breed. These remarkable long-distance fliers include species like the Eastern Wood Pewee (*Contopus virens*), Great-crested Flycatcher (*Myiarchus crinitus*), Orchard Oriole (*Icterus spurius*), Blue Grosbeak (*Passerina caerulea*), Summer Tanager (*Piranga rubra*), and Indigo Bunting (*Passerina cyanea*).

Orchard Oriole.

Summer Tanager.

Indigo Bunting.

Reptiles and Amphibians

Eastern Indigo Snake

Drymarchon corais couperi

These exceptionally rare creatures are no longer a common species in longleaf pine forests. Though they were historically found from southern South Carolina to Mississippi, today they are found only in Georgia and Florida. Experimental reintroductions are being conducted in Conecuh National Forest in Alabama and Apalachicola Bluffs and Ravines Preserve in Florida. Reproduction in the wild has been documented in the Conecuh National Forest population. Time will tell the success of this effort. Eastern indigo snakes have huge home ranges (250 acres or more) and are therefore susceptible to habitat fragmentation and loss. Combined with historical collections for the pet trade, their numbers have declined significantly over the years.

SIZE: This snake grows up to 8½ feet in length, making it the largest snake in North America.

Eastern indigo snake.

DESCRIPTION: Lustrous blue-black color. The sides of the head, chin, and throat area are suffused with orange/red.

BEHAVIOR: This snake is nonvenomous yet is a fierce predator in longleaf pine forests. It feeds on birds, small mammals, frogs, small turtles, fish, and snakes (including venomous ones such as the eastern diamondback rattlesnake). Prey is usually eaten alive after being immobilized by a crushing bite from its jaws (it does not constrict like a python). In the colder months in the northern part of its range, this snake will spend most of its time on upland sandhill ridges and use gopher tortoise burrows extensively. In the warmer months, these snakes can be found foraging the edges of swamps scattered within the longleaf pine landscape.

ADDITIONAL NOTES: Though unusually docile with humans, when aggravated this snake will hiss, vibrate its tail, and flatten its neck to appear threatening.

Pine Snakes

Pine Snake (*Pituophis melanoleucus*)
Louisiana Pine Snake (*Pituophis ruthveni*)

These large snakes spend most of their time underground but are occasionally seen as they move about during the daytime in search of food. They prefer sandy soils and can often be found in pursuit of small rodents, especially those found in the side root channels of burned-out stumps. Two species and multiple subspecies of pine snake can be found throughout the range of longleaf pine.

SIZE: Its robust body has an average length of 4–5½ feet.

DESCRIPTION: The different species and subspecies of pine snakes have varying patterns and coloration, but some may be difficult to tell apart in the field. The black pine snake (*Pituophis melanoleucus lodingi*) is a uniform black or dark brown, whereas the Florida pine snake (*Pituophis melanoleucus mugitus*) may be white, yellow, or light gray or buff with dark brown blotches. Northern pine snakes (*Pituophis melanoleucus melanoleucus*) have a light background with dark contrasting patches. Louisiana pine snakes have a diffuse pattern toward the head; this pattern gradually becomes more organized

Louisiana pine snake.

Black pine snake.

Northern pine snake.

down the body and tail. In some areas, geographic distribution can be the most useful way to distinguish the species or subspecies.

BEHAVIOR: These snakes will typically rely on constriction to kill their small prey, such as cotton rats, pocket gophers, and ground birds. When this snake feels threatened it will coil, rear back, produce a loud hissing noise, and strike, repeatedly mimicking the behavior and sound of a rattlesnake. The snakes also have an enlarged rostral scale that they use to root through loose soil in search of prey. Louisiana pine snakes rely heavily on Baird's pocket gopher (*Geomys breviceps*) for tunnels and as prey.

Florida pine snake.

ADDITIONAL NOTES: Pine snakes are further divided into three subspecies: black pine snake (*Pituophis melanoleucus lodingi*), northern pine snake (*Pituophis melanoleucus melanoleucus*), and the Florida pine snake (*Pituophis melanoleucus mugitus*).

Eastern Coral Snake

Micrurus fulvius

These snakes are reclusive and difficult to observe. However, they have been seen using a variety of refugia, from gopher tortoise burrows to stump holes and underneath the loosened bark of longleaf pine stumps.

SIZE: The slender eastern coral snake averages 20–30 inches long.

DESCRIPTION: Bright alternating bands of red, yellow, and black run the length of its body.

BEHAVIOR: This snake's specialized diet includes smaller snakes, glass lizards, and skinks that it injects with venom.

ADDITIONAL NOTES: The venom of the eastern coral snake is similar to that of cobras. This snake does not have rattles or other auditory cues to warn potential predators. Instead, it relies on bright coloration to broadcast "stay away, I'm dangerous." The red, black, and yellow of the coral snake are similar to the nonvenomous scarlet kingsnake (*Lampropeltis triangulum elapsoides*) and scarlet snake (*Cemophora coccinea*). You can tell the three snakes apart by looking at their colored bands and remembering this mnemonic: "black touches red, kills you dead; black touches yellow, nice fellow."

Eastern coral snake.

Scarlet kingsnake for comparison to Eastern coral snake.

Pinewoods snake.

Pinewoods Snake

Rhadinaea flavilata

These are found sporadically throughout longleaf pine's range. They are small, mostly fossorial, and very secretive. They are restricted to piney woods in the coastal plain of the Southeast.

SIZE: The snakes are small; the maximum length of adults is around 1 foot.

DESCRIPTION: Pinewoods snakes range from golden brown to reddish brown. A dark stripe runs through the eye. The specific descriptive name *flavilata* translates to "yellow lip," referring to the yellowish coloration of upper lip scales.

BEHAVIOR: They have a mild venom in their saliva that is used to subdue small prey such as insects, salamanders, and small lizards. They are not harmful to humans. They lay one to four eggs in midsummer under rotten wood.

ADDITIONAL NOTES: They can be differentiated from other small brown snakes by the dark stripe across the eye and yellow lip scales.

Short-tailed Snake

Lampropeltis extenuata

These small, slender, fossorial snakes of pine flatwoods are seldom seen. They are named for their short tails that make up a small percentage of their length.

SIZE: They reach a maximum length of about 20 inches.

DESCRIPTION: These snakes have gray bodies with dark spots along their length and smooth scales. The spots may be separated by red or yellow. The head of this snake is indistinct from the rest of the body.

BEHAVIOR: Short-tailed snakes spend the majority of their time underground in sandy dry soils. They feed primarily on small vertebrates, especially smaller snakes. Little is known about how they reproduce except that they lay eggs.

ADDITIONAL NOTES: The pattern and extremely short tail distinguish these from other kingsnakes. They live exclusively in longleaf pine and oak scrub habitats of the central region of peninsular Florida. Their pattern is similar to that of both a pigmy rattlesnake (*Sistrurus miliarius*) and southern hognose snake (*Heterodon simus*).

Short-tailed snake.

Southeastern Crowned Snake

Tantilla coronata

These small snakes of dry habitats occur across most of the range of longleaf pine.

SIZE: Adults reach a maximum length of about 10 inches. The hatchlings are about 3 inches long.

DESCRIPTION: They are khaki brown with a small dark crown or ring at the base of the head. Except for their size, hatchlings look almost identical to the adults. The back is usually lighter than the undersides. Snakes appear shiny due to the smooth scales and are mildly iridescent.

BEHAVIOR: These snakes are mostly fossorial and secretive. They are seldom seen by casual observers even when common. Southeastern crowned snakes are centipede specialists but will consume other small arthropods. They are mildly venomous but pose no threat to humans.

ADDITIONAL NOTES: The small dark ring around the base of the head distinguishes this snake from other small brown snakes in the area. Various subspecies can be found from peninsular Florida all the way up to montane longleaf pine habitats of Alabama and Georgia.

Southeastern crowned snake.

Eastern Diamondback Rattlesnake

Crotalus adamanteus

These snakes are found only in the coastal plain and not in mountain longleaf pine habitat. Their numbers have declined over recent years due to a number of factors, including persecution, habitat loss, and vehicle mortality. In parts of their range, they can still be found crossing forest roads in the early evening hours in the warmer summer months or lounging near the mouth of a gopher tortoise burrow when the days are cool and short.

SIZE: This is the largest rattlesnake species found anywhere in the world. Typical adults are 3–6 feet long and can weigh up to 10 pounds. Exceptional individuals have been recorded at almost 8 feet long. Males are generally larger than females.

DESCRIPTION: Their diamond pattern of brown, black, tan, and yellow can make this snake nearly invisible in leaf litter and dappled sunlight.

Eastern diamondback rattlesnake.

BEHAVIOR: Rattlesnakes are highly adapted ambush predators. With their superb camouflage they coil under small pine seedlings or palmetto bushes, waiting for unsuspecting prey to pass. They feed mostly on rabbits, squirrels, and cotton rats. Rattlesnakes are pit vipers and possess two heat-sensitive pits on either side of their face. These organs detect body heat and aid in locating prey and increasing strike accuracy. The large folding fangs swing into position during a strike and inject a powerful venom into their prey. This venom both kills the prey and helps in digestion. Females give birth to live young through a process called ovoviviparity.

ADDITIONAL NOTES: These snakes have been hunted ruthlessly over many decades and exploited for rattlesnake roundups. People who work in the woods are at a far greater risk of bee stings, ant stings, and falling tree limbs than of being bitten by a rattlesnake. Though the sounds of a rattling snake can strike fear in the hearts of many people, this telltale rattle evolved for a reason: to alert those around it not to mess with it. They simply want to stand their ground until they can safely retreat.

Timber Rattlesnake

Crotalus horridus

Though these snakes are found along the entire eastern half of the United States, they occur throughout longleaf habitats except for those in peninsular Florida and the extreme southern portions of the Gulf States. This is the Southeast's widest ranging rattlesnake and is locally called canebrake rattlesnake due to its use of historic dense stands of river cane (*Arundinaria* spp.).

SIZE: Adults generally reach maximum lengths of about 5 feet, although larger individuals have been recorded. Newborns are less than 1 foot long.

DESCRIPTION: Timber rattlesnakes generally range from tan, yellow, and pink to black, with dark chevron patterns extending down their back. Their tail is a deep, dark black terminating in rattles made of specialized scales. A pit viper, the snake has heat-sensing organs between the mouth nostrils that allow it to sense prey.

BEHAVIOR: These snakes are ambush predators that lie and wait for prey to come close before delivering a fatal strike. Newborns and young snakes may eat a variety of insects. Adults specialize in small mammals and may

Timber rattlesnake.

occasionally feed on other vertebrates. Timber rattlesnakes rely on their camouflage to remain hidden and unnoticed. When this fails, they may rattle as a warning but rarely bite in defense. Breeding takes place when snakes are approximately 7–10 years old and may only occur every 2 years. Females give birth to live young through ovoviviparity.

ADDITIONAL NOTES: Timber rattlesnakes overlap with eastern diamondback rattlesnakes in parts of the range of longleaf pine. The chevron pattern extending down the back of the timber rattlesnake is distinctive. This snake's venom is toxic to humans, and any bite should be treated as a medical emergency. Like other eastern rattlesnakes, timber rattlesnakes are in decline throughout most of their range.

Pigmy Rattlesnake

Sistrurus miliarius

These small venomous pit vipers occur throughout longleaf pine's range. This species is much smaller than other rattlesnake species found in this forest type. They also lack the large heads seen in the region's *Crotalus* species. Pigmy rattlesnakes are rather distinct from the other two rattlesnake species

Pigmy rattlesnake.

that occur in longleaf pine ecosystems. These may also be called ground rattlers.

SIZE: As the name implies, pigmy rattlesnakes are small snakes that reach a maximum length of about 3 feet. Neonates are around 6 inches. Females give birth to live young through ovoviviparity.

DESCRIPTION: These snakes are highly variable in color, with the background ranging from gray, tan, and pink to red. Dusky and western pigmy rattlesnakes tend toward duller gray and tan, while Carolina pigmy rattlesnakes can be brilliant red. All three subspecies have dark patches across the entire body. The smaller size and irregular blotches separate the other species that have distinct, diamond-shaped (eastern diamondback rattlesnake) or chevron-shaped (timber rattlesnake) patterns down the back. The tails terminate in tiny rattles that usually consist of only a couple segments.

BEHAVIOR: Adults and young both eat a variety of insects. Adults will also eat a variety of small vertebrates, including small rodents, lizards, and frogs. They rely on camouflage to remain hidden so they can ambush their prey with a venomous bite. They also rely on their camouflage to remain unnoticed by attackers, since their rattles are so small and seldom heard. Instead of the rattle warning of the larger diamondback and timber rattlesnakes, pigmy rattlesnakes often twitch their heads from side to side to warn those who get too close. Bites are a last-resort defense, as venom takes time and energy to produce. Reproduction occurs in the fall. Like other pit vipers, females give birth to live young through ovoviviparity.

ADDITIONAL NOTES: Three subspecies occur: the Carolina pigmy rattlesnake (*Sistrurus miliarius miliarius*), the dusky pigmy rattlesnake (*Sistrurus miliarius barbouri*), and the western pigmy rattlesnake (*Sistrurus miliarius streckeri*).

Eastern Racers

Northern black racer (*Coluber constrictor constrictor*)
Southern black racer (*Coluber constrictor priapus*)

These are both medium-sized snakes. They are black and fast. Usually, you are just able to catch a glimpse of them before they flee. They are long and slender with narrow heads. Several subspecies of eastern racer occur within longleaf pine forests, but northern and southern black racers are the predominant subspecies encountered throughout most of longleaf pine's range.

SIZE: Adults can reach 5 feet long, while hatchlings are usually under 8 inches.

DESCRIPTION: Adult southern and northern racers are black on the dorsal surface and dark gray on the ventral surface. Most individuals have a white chin and neck. Neonates and young juveniles look quite different from adults. Individuals smaller than 1 foot in length have a tan or gray background with dark splotches running the length of the body.

BEHAVIOR: Adults and young eat any invertebrates and vertebrates that they are able to overpower, including other snakes. The constrictor designation is a misnomer, as these snakes do not constrict prey. Instead, they rely on overpowering the prey and swallowing it alive. They are most active during the day and are often found in the open even on the region's warmest days. These snakes do not readily climb and are generally found on the ground.

Northern black racer.

Southern black racer.

Snakes will often flee quickly toward cover in the direction they are facing, which sometimes means they come toward the perceived threat.

ADDITIONAL NOTES: These snakes are often mistaken for other black snakes that occur within longleaf pine communities, including eastern indigo snakes, rat snakes, black pine snakes, and eastern kingsnakes (*Lampropelits getula*). They can be differentiated from these species by a thinner streamlined body, light chins with dark underbelly, and lack of any white patterning on the upper surface of the body.

Corn Snake

Pantherophis guttatus

These medium-sized snakes are present in most of the range of longleaf pine. They are highly variable in coloration, and some color variants are specific to certain regions. Those that are red often are called red rat snakes.

SIZE: Adults can reach about 6 feet long. Hatchlings are usually less than 12 inches.

DESCRIPTION: They are patterned with alternating patches of light gray, white, yellow, brilliant orange-red, rusty red, and black that extend down the back and sides. The belly is a checkerboard of brilliant white and black. In some parts of their range they can look very much like rat snakes but are distinguished by the spear-point pattern on the top of the head.

BEHAVIOR: Adults and young both eat a variety of invertebrates and vertebrates that they constrict before eating. In addition, corn snakes are known to eat a variety of bird eggs, especially ground and cavity nesting birds. These snakes can be active both day or night depending on temperature. They spend a significant amount of time under cover on the ground and also off the ground in shrubs/trees. Eggs are laid once a year in midsummer.

Corn snake.

ADDITIONAL NOTES: Corn snakes are one of the most popular pet snakes in the world. Hundreds of color variations have been created through artificial selection, and many variations have been found in the wild and subsequently commercialized. Their popularity comes from their docile nature, reluctance to bite, manageable size, and adaptability. Even wild corn snakes are usually quite tame when handled for research purposes.

Hognose Snakes

Eastern Hognose Snake (*Heterodon platirhinos*)
Southern Hognose Snakes (*Heterodon simus*)

Hognose snakes are small snakes that are familiar occupants of longleaf pine forests. Both species are characterized by distinct scales that give them an upturned snout.

SIZE: Eastern hognose snakes can reach 4½ feet long, while the smaller southern hognose reaches a maximum of around 2 feet.

DESCRIPTION: Both species are variable in their patterning, but eastern hognose snakes show the greatest color variation. Adults of both species are often gray or tan with contrasting dark blotches. Both species may have some diffuse orange coloration throughout, but this may be much more prominent in eastern hognose. Eastern hognose snakes may also be completely gray or

Melanistic eastern hognose snake.

black, with the pattern obscure or hidden. Hatchlings are often more boldly patterned than adults.

BEHAVIOR: Both of these snakes specialize in eating toads. They are mildly venomous, with small rear fangs. The fangs are used to deliver venom as well as to hold onto toads until the venom takes effect. Hognose snakes rarely bite, and even when they do, the venom is seldom problematic for humans. When threatened, hognose snakes have dramatic and predictable displays. The first display is meant to intimidate: they spread the flanks of their necks, making themselves appear larger while breathing in and exhaling loudly to produce a hissing noise. If that fails, they play dead by rolling over, exuding foul-smelling excrement, and laying with their mouth partially opened, complete with tongue lolling out. Even when rolled over, they will immediately resume the position and display. A last resort is to vomit their last toad meal.

ADDITIONAL NOTES: While both snakes appear similar, eastern hognose snakes have lighter coloration on the underside of the tail (from vent to tip)

Eastern hognose snake playing dead.

Southern hognose snake.

compared to the rest of the body. Southern hognose snakes have similar coloration throughout. Southern hognose snakes have experienced significant declines throughout their range, due to habitat loss and introduction of invasive species.

Eastern Kingsnake

Lampropeltis getula

These large snakes have strong bodies and heads that are of similar width to the rest of their body, making it appear that they have no neck. There are three large plate scales on the top of the head.

SIZE: Adults can reach 4 feet long. Hatchlings are usually under 12 inches.

DESCRIPTION: Coloration even within subspecies can be highly variable. Speckled kingsnakes are generally dark in coloration, with varying density of white or yellow spots across the entirety of their body. The eastern subspecies is generally dark, with a white or yellow chain pattern. Some speckled

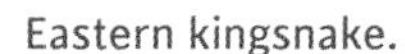

individuals occur on North Carolina's Outer Banks and in the southern Appalachians.

BEHAVIOR: They eat a variety of prey, including small mammals, birds, and amphibians. Most notably, eastern kingsnakes are known for their tendency to eat other snakes. They are resistant to local pit viper venoms, earning them the name kingsnake. Eggs are laid in early summer, and hatchlings emerge in late summer to fall.

ADDITIONAL NOTES: There are several subspecies throughout the range of longleaf pine. Speckled kingsnake (*Lampropeltis getula holbrooki*) and the eastern subspecies (*Lampropeltis getula getula*) are the most common and widespread throughout this region.

Eastern Coachwhip

Masticophis flagellum

This streamlined and fast snake is named for the similarity of its scale pattern to a braided whip. Found throughout most of longleaf pine's range, this species relies on open forest stands.

SIZE: Though typically 5–6 feet in length, they are one of the longest snakes in this forest type, with rare sightings of snakes up to 8½ feet long.

DESCRIPTION: These long, slender snakes have distinct patterning. Adults are unique in that the head starts off black but fades into a tan braided pattern farther down the body. The ratio of black to tan varies across the longleaf pine range, and it is not unusual to find entirely tan individuals in some areas. Hatchlings and young snakes are quite different from adults, having a blotched pattern on a light background.

BEHAVIOR: They do not constrict their prey, but instead rely on their extreme speed and size to overcome prey and swallow them alive. Eastern coachwhips are also known for a behavior known as periscoping, in which they hold a significant portion of the body upright and turn the head from side to side to look for prey or predators above the grasses and herbaceous layer.

ADDITIONAL NOTES: These are generally the fastest snakes in the longleaf pine forest and, for such a large snake, can simply disappear into the

Eastern coachwhip.

undergrowth in the blink of an eye. They are very active predators with excellent vision.

Copperhead

Agkistrodon contortrix

These well-camouflaged, venomous pit vipers have a wide range that includes longleaf pine forests except in most of Florida. The multiple subspecies in this region are not significantly different to the casual observer.

SIZE: Adults can reach a maximum 5 feet long but are on average 2–3 feet. Young snakes are generally under 1 foot.

DESCRIPTION: Copperheads have a gray to light-brown background, with cinnamon-brown hourglass crossbands. Their triangular heads have heat-sensing pits and catlike pupils. Mature snakes often have large muscular bodies compared to the length; the young have a sulfur-yellow tail.

BEHAVIOR: Adults eat a variety of insects and vertebrates, with small rodents making up their primary diet. Young copperheads use their bright

Copperhead.

sulfur-colored tail to lure potential prey. They are ambush predators that inject venom into their prey before tracking down the now-paralyzed victim and swallowing it whole. The young are born through ovoviviparity.

ADDITIONAL NOTES: Copperheads have a distinct pattern that separates them from other snakes. While some nonvenomous snakes can have similar color, the patterns of species like northern watersnake (*Nerodia sipedon*) or reddish variety of the banded watersnake (*Nerodia fasciata*) lack the cinnamon-brown hourglass crossbands.

Cottonmouth

Agkistrodon piscivorus

These large-bodied snakes with dark coloration are often found in or near wetlands and water bodies. They are often common entities of wetlands within longleaf pine communities throughout the tree's range. These snakes are sometimes called water moccasins.

SIZE: Adults can reach a maximum of 6 feet long. Young snakes are generally under 1 foot long.

Cottonmouth.

DESCRIPTION: Adults have an indistinct crossband pattern over a dark brown background. Young cottonmouths have a pattern very similar to copperhead snakes but are often described as pixelated. Some snakes, especially younger or recently shed snakes, may have hourglass saddles like copperheads. Young snakes are more boldly patterned with generally brighter colors. They are named for the bright white interior of the mouth.

BEHAVIOR: Cottonmouths spend much of their time near water. Their diet consists of a variety of vertebrates, including frogs, fish, rodents, and other snakes. As ambush predators, they lie hidden and strike to inject venom into unsuspecting prey. The prey then succumbs to the venom before being tracked and swallowed whole. The venom of cottonmouths is very potent; their bites are dangerous for humans. These snakes exhibit an elaborate open-mouth defense display. This display gives them the name of cottonmouth.

ADDITIONAL NOTES: Northern and banded watersnakes can have similar coloration but lack the banding. In addition, cottonmouths often have much more robust bodies. The open-mouth display of cottonmouths is sometimes a defining characteristic, though the old stories of cottonmouths chasing humans is false.

Mole Skink

Plestiodon egregious

These small, slender, semi-fossorial lizards occur in Georgia, Alabama, and Florida, possibly even extending into South Carolina. These lizards often slither rather than run for cover, and observers often get only a brief glimpse before the lizard has made it safely under cover. They rarely climb trees and thus can be killed by frequent fires in longleaf pine forests. They also burrow into disturbed sand such as pocket gopher mounds and gopher tortoise burrow aprons.

SIZE: Adults are 3½–6 inches long, and hatchlings are around 1 inch. Tails can make up a large portion of the lizard's length, and since these lizards are capable of autotomy (shedding the tail), length should be measured snout to vent.

DESCRIPTION: Mole skinks are generally grayish to brown or bronze with a bright red tail that does not fade with age. They have two light stripes running the length of the body. The red on the tail is distinctive. In some subspecies, the red tail will fade to blue starting at roughly half the tail's length.

BEHAVIOR: They are highly secretive. Little is known about their life cycle. Females lay up to 6 eggs in moist soil and guard the nest during the winter. Adults hunt and feed in dry open areas both above and below ground, particularly in longleaf pine and associated scrub habitats for the northern subspecies. Mole skinks feed mostly on small insects.

Mole skink.

ADDITIONAL NOTES: Several subspecies exist: Florida Keys mole skink (*Plestiodon egregius egregius*), Cedar Key mole skink (*Plestiodon egregius insularis*), blue-tailed mole skink (*Plestiodon egregius lividus*), peninsula mole skink (*Plestiodon egregius onocrepis*), and northern mole skink (*Plestiodon egregius similis*). Some subspecies are state or federally protected.

Lined Skinks

Five-lined Skink (*Plestiodon fasciatus*)
Southeastern Five-lined Skink (*Plestiodon inexpectatus*)
Coal Skink (*Plestiodon anthracinus*)
Broadhead Skink (*Plestiodon laticeps*)

Generally similar in appearance, these four skinks may occur within longleaf pine forests. All four are locally known as blue-tailed skinks and, except for coal skink, can be difficult to distinguish in the field.

SIZE: Five-lined skinks, coal skinks, and southeastern five-lined skinks are similar in length, with five-lined skinks being slightly larger at 6–8 inches. Adult broadhead skinks are perhaps the largest of the native lizards typical

Five-lined skink.

Southeastern five-lined skink.

Coal skink.

Male broadhead skink.

Female broadhead skink.

to this region, with only a few of the glass lizards (like the eastern glass lizard [*Ophisaurus ventralis*]) reaching a larger body weight and length. Adult broadhead skinks can reach 6–10 inches in length. Tail length is variable, as all four species are capable of autotomy, so length should be measured snout to vent.

DESCRIPTION: Juveniles and females of these species have similar color characteristics. The basic color of juveniles is dark overall, with lines running down the length of the body and a bright blue tail. Coal skinks have a lighter upper surface than the rest of its body and are missing the fifth line above the backbone seen in the other three species. Its sides are flanked by darker coloration than the top. The other three species have a dark background with five lines running longitudinally down the body. As they age, even the females may lose the blue tail seen in juveniles. The lines in males will gradually fade and mostly disappear altogether in broadhead skinks. Males of all four species can develop red on their head. Broadhead skinks develop particularly broad cheek muscles that give them their name.

BEHAVIOR: All four species eat a variety of insects. Eggs are laid under logs or leaves in shallow chambers that the females guard. Males can be territorial, and broadhead skinks especially defend areas against rival males. Coal skinks spend much of their time on the ground or just above the surface on rocks or downed woody debris. The other three species will venture up into decaying snags in search of food. All species can be found basking in sunny woodland openings.

ADDITIONAL NOTES: A popular myth among locals has it that these species can sting with their blue tails, leading them to be called scorpions in some areas. Distinguishing between five-lined, southeastern five-lined, and broadhead skinks relies on careful examination of the scales on the tails and lips. Five-lined skinks have a row of enlarged scales underneath the tail, distinguishing them from southeastern five-lined skinks, and four lip scales, compared to five on the broadhead.

Ground Skink

Scincella lateralis

This tiniest of skinks can be found in the leaf litter of many different forest types. The ground skink can be especially abundant in longleaf pine forests and associated oak-scrub areas.

SIZE: Adults can reach a maximum 5 inches long, while the hatchlings are under 1 inch. These are the smallest lizards encountered in longleaf pine communities. Ground skinks are capable of autotomy, so use the snout-to-vent measurement.

DESCRIPTION: They are bronze on the dorsal surface and have a dark stripe running down each side of the body. The ventral surface is usually white to yellow, with no distinct markings. Ground skinks are most similar to mole

Ground skink.

skinks in both behavior and appearance, but ground skinks do not have the red tail.

BEHAVIOR: These skinks flee from approaching predators by burying into leaf litter in a slithering motion without much assistance from the legs. They forage among leaf litter and detritus on the forest floor, rarely climbing more than 1 foot off the ground. Prey consists of small arthropods and other invertebrates. Up to 5 eggs are laid in rotten logs or other moist organic material, including small soil mounds. Females may be able to lay more than one clutch each year.

ADDITIONAL NOTES: The plain color of the tail is the most distinguishing characteristic compared to other skink species within longleaf pine forests.

Eastern Fence Lizard

Sceloporus undulatus

These lizards are a common sight in longleaf pine forests. They can frequently be seen (or heard) moving quickly up the side of longleaf pine trees. They are also known as either fence swift lizards or pine lizards.

SIZE: This small lizard typically measures up to 7½ inches snout to vent.

DESCRIPTION: They have a roughly scaled body that is typically light gray, with a dark chevron-like pattern along the back. Males have blue patches on

Eastern fence lizard.

the belly and throat. In burned-over areas, they are quite concealed among the ash on the ground and on charred tree trunks.

BEHAVIOR: These lizards rarely chase prey. Instead, they typically wait for crickets, grasshoppers, beetles, or centipedes to come to them. They sometimes forage for ants entering and exiting their mounds.

ADDITIONAL NOTES: As with many of this region's lizards, the tails of eastern fence lizards will fall off if grabbed. The broken tail fragment will continue to wiggle, distracting the would-be predator as the tailless lizard escapes.

Eastern Glass Lizard

Ophisaurus ventralis

These are often mistaken for snakes. While they have no legs, they have eyelids, ear openings, and a fixed jaw that separate them from snakes. In addition, like other lizards, glass lizards can shed their tail.

SIZE: The hatchlings are generally 6–8 inches in length. Adults can reach 40 inches long head to tail. Tail length is highly variable and may make up over half of the total length, so length should be measured snout to vent.

DESCRIPTION: The body of the adult is generally heavily speckled on the dorsal surface and golden on the ventral surface, separated by a groove running the length of the side. The young have two stripes running longitudinally down the back, with an overall tan coloration.

BEHAVIOR: These lizards occur in open habitats with generally grassy or dense understory herbaceous layers. All glass lizards have fixed jaws, limiting the size of the prey that can be swallowed. Because of this, these lizards consume mostly small insects. When threatened or handled roughly, their tail may "shatter" into one or more pieces, thus the name glass lizard. Eastern glass lizards should be handled as little as possible to avoid causing unintentional harm. Severed tails will often move for several minutes to distract predators. Tails contain a significant portion of energy stores, so losing them can adversely affect the individual.

ADDITIONAL NOTES: Three other glass lizards occur within certain parts of the longleaf pine range: slender glass lizard (*Ophisaurus attenuatus*), island glass lizard (*Ophisaurus compressus*), and mimic glass lizard (*Ophisaurus mimicus*). Eastern glass lizards are often the most common and largest of the glass lizard species and can be distinguished from the others by lacking lengthwise dark stripes below the lateral groove or tail.

Eastern glass lizard.

Green Anole
Anolis carolinensis

Green anole.

These small lizards are common throughout longleaf pine's range. Green anoles are good climbers and can be found near the ground all the way up into tree canopies. Locals may also incorrectly call these chameleons.

SIZE: Since these lizards are capable of autotomy, length should be measured snout to vent. Adult green anoles are about 5–8 inches long. New hatchlings from eggs can be around 2 inches long. Hatchling and juvenile anoles look exactly like the adults except size.

DESCRIPTION: They can change body color from light brown/gray to brilliant green. This color change is rapid in comparison to the region's other native reptiles. Color is generally dependent on the body temperature and mood of the individual; stressed individuals will often appear brownish.

BEHAVIOR: These lizards are incredibly adaptable and can be found throughout many habitats in the southeastern United States. They are especially abundant in many longleaf pine forests, particularly around shrubby understories associated with wetlands. Green anoles are found on trees and may be the only lizard encountered in the longleaf pine canopy. Males have a bright pinkish-red dewlap (fold of skin) below the chin. During territorial displays this dewlap is extended and accompanied by multiple head bobs to warn intruders as well as attract mates. Males can be extremely territorial and aggressively defend their territory from rivals. Green anoles eat mostly small arthropods, although they have been known to consume smaller lizards.

ADDITIONAL NOTES: The nonnative brown anoles (*Anolis sagrei*) also occur within the range of longleaf pine. Originally from Cuba and the Bahamas, they are brown and often have a mottled dark pattern down the back. They also have a blunter snout. Males of the brown anole have a yellow-orange-red dewlap rather than the pink-red seen in the native green anole. Introduction of brown anoles has modified the behavior of native green anoles, often leading them to stay in tree canopies and avoid competition with brown anoles, which forage closer to the ground.

Six-Lined Racerunner

Cnemidophorus sexlineatus

These small, colorful lizards move with incredible speed. In some longleaf pine forests, they may be the most common reptile species encountered.

SIZE: Adults can reach about 9 inches long, while hatchlings are usually under 3 inches. Six-lined racerunners are capable of autotomy, so measure the length snout to vent.

DESCRIPTION: They have a dark background similar to the five-lined skink group; however, they are matte rather than glossy and have 6 yellow lines running down the length of the body. Adult males can develop blue under the chin and sky-blue tails with a blue to yellow ventral surface. Males tend to be brighter overall, especially during breeding season. Females have a white belly with dull yellow highlights all year round.

BEHAVIOR: These highly active predators eat a variety of arthropods. Adults and young are extremely quick and will dart away if encountered. They are often very active even in extreme heat and prefer open areas with grassy and herbaceous understory or bare soil. In longleaf pine ecosystems, these lizards can reach high densities, especially since they do not defend territories. Six-lined racerunners almost never leave the ground.

ADDITIONAL NOTES: While similar to five-lined, southeastern five-lined, and broad-head skinks, six-lined racerunners have small, grainy dorsal scales

Six-lined racerunner.

and larger square scales on the tail and belly. In addition, six-lined racerunners have large claws relative to body size, and their speed and ground-dwelling habit make them almost unmistakable.

Box Turtle

Terrapene carolina

This is one of two land turtles found in longleaf pine forests, the other being the gopher tortoise. However, unlike the gopher tortoise, box turtles are the pond turtle family, Emydidae. These relatively slow-moving animals can frequently be seen shuffling through the forest litter, usually along the ecotones between uplands and wet seep areas.

SIZE: This small turtle is usually 4–8½ inches long.

DESCRIPTION: Males generally have a concave plastron, are more colorful than females, and have more flared shell margins, larger curved claws on

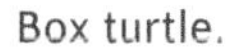

Box turtle.

their hind feet, and red eyes. Females have a flat plastron, duller coloration, more rounded shells, and shorter straight claws on the hind feet, and their eyes are yellowish brown. In general, the carapace is dark brown, with a pattern of yellow or orange radiating lines or spots.

BEHAVIOR: They slowly walk through the leafy underbrush in search of insects, fruits, mushrooms, and occasionally carrion. When alarmed, these turtles can close up in their hinged shell completely to protect themselves, earning them the name of box turtle.

ADDITIONAL NOTES: These turtles are unlikely to outrun the flames of the frequent fires in longleaf pine forests; therefore, it is not unusual to see box turtles with burn scars on their shells. Their shell has the remarkable ability to heal, but the regrown shell loses its regular and colorful patterning, forever marking their ordeal. More often they rely on wet areas, slight depressions in the ground, and gopher tortoise burrows as refuge from passing flames. Four subspecies of box turtle are found across the longleaf pine range: eastern box turtle (*Terrapene carolina carolina*), Florida box turtle (*Terrapene carolina bauri*), Gulf Coast box turtle (*Terrapene carolina major*), and three-toed box turtle (*Terrapene carolina triunguis*).

Gopher Tortoise

Gopherus polyphemus

These tortoises dig unusually long burrows using their strong, shovel-like front feet. Such burrows are typically long and straight but can be branched, corkscrewed, shallow, or deep depending on soil texture and underground obstacles. The typical burrow is 15–20 feet long, but one burrow was recorded at 47½ feet. The back of the burrow may be 10–20 feet below the surface. The burrow allows tortoises to escape from periodic fires, regulate their body temperature, and conserve water. Numerous animals and arthropods use these burrows as a year-round residence, while others use them only temporarily. In fact, over 300 species of animals depend on the gopher tortoise directly or indirectly. For this reason, the gopher tortoise is considered a keystone species of longleaf pine forests. They are restricted to sandy uplands of the coastal plain from southeastern South Carolina into eastern Louisiana.

Adult gopher tortoise in front of its burrow.

SIZE: Females are typically larger than males. Adults are about 9¼–15 inches long, with some individuals weighing up to 29 pounds.

DESCRIPTION: Typically, this tortoise is plain dark tan or gray.

BEHAVIOR: Gopher tortoises are often found basking on the apron of the burrow. They are plant eaters, and during the warmer months they can be found feeding on foliage and fruits of many plants in the longleaf pine forest. They are also important seed dispersers for many plants that they consume as the seeds pass through their gut.

ADDITIONAL NOTES: First described by science in 1802, the species is named after Polyphemus, the legendary cave-dwelling giant in Homer's

Odyssey. Though protected today by state or federal laws, gopher tortoises were once considered a food staple by many pioneering country folks. Men often "pulled" (fished) gopher tortoises (also called Hoover chickens, cracker chicken, or Florida bacon).

Flatwoods Salamanders

Frosted Flatwoods Salamander (*Ambystoma cingulatum*)
Reticulated Flatwoods Salamander (*Ambystoma bishopi*)

These animals are unique in that they are found in pine forests yet also require adjacent wetlands. They require flat, low-lying pine forests with interspersed isolated wetlands. They breed in small, shallow ponds typically characterized by a sparse canopy of pond cypress (*Taxodium ascendens*) or black tupelo (*Nyssa biflora*) and dominated by abundant herbaceous vegetation. Both species are now very rare within their ranges and are an indicator of high-quality longleaf pine flatwoods or savannas. The flatwoods salamanders on either side of the Apalachicola River in Florida are separate species: reticulated and frosted.

SIZE: It has a stubby head and limbs on a body typically 3½–5 inches long. The flatwoods salamander would look slightly larger than the reticulated salamander.

Frosted flatwoods salamander.

Reticulated flatwoods salamander.

DESCRIPTION: Both species are dark with gray netlike markings on their back. The belly is black, with gray-to-white specks. The reticulated salamander is described more as salt and pepper in its coloring, while the flatwoods salamander has white spots on black to chocolate black.

BEHAVIOR: The animal is fairly solitary, spending the majority of its time underground in cool, damp spaces such as under rocks, logs, or crawfish burrows during daylight hours. At night they emerge to feed on small worms, beetle larvae, and termites.

ADDITIONAL NOTES: Their breeding wetlands must dry out seasonally so that predatory fish cannot become established. Fish will eat the eggs and larvae of flatwoods salamanders.

Eastern Tiger Salamander

Ambystoma tigrinum

These are the largest terrestrial salamanders in longleaf pine forests. They are often secretive, but conspicuous when they are seen. Like other mole salamanders, it requires two habitats: ponds for breeding and upland habitat such as sandhills or longleaf pine flatwoods during other seasons.

SIZE: Adults can reach 1 foot long. Hatchlings are often under 1 inch, while recently metamorphosed salamanders are generally under 5 inches. Adult tiger salamanders are the largest terrestrial salamanders in the eastern United States.

DESCRIPTION: The adult is dark-bodied, often black, with yellow spots/marbling. A member of the mole salamander family, it has a large head with conspicuous eyes and wide mouth. Larval salamanders have exposed gills and have similar, but muted, patterning to the adults.

BEHAVIOR: Adults eat a wide variety of insects, amphibians, reptiles, and occasionally small mammals. They breed in the winter, and breeding takes place in large groups. Metamorphosis is completed within 10 weeks to 6 months. Some adults may never complete metamorphosis and are referred to as paedomorphic adults, retaining larval salamander characteristics their entire life and never leaving the water.

Eastern tiger salamander.

ADDITIONAL NOTES: These salamanders have a distinct pattern but most closely resemble spotted salamanders (*Ambystoma maculatum*). Their spots/mottling are less uniform than spotted salamanders and often a much duller yellow.

Dwarf Salamander

Eurycea quadridigitata complex

Once recognized as one species across the longleaf pine range, dwarf salamanders are now split into multiple species, including the dwarf salamander (*Eurycea quadridigitata*) found in southeast Georgia, peninsular Florida, and coastal South and North Carolina; the Chamberlain's dwarf salamander (*Eurycea chamberlaini*), found in interior South and North Carolina; the Hillis's dwarf salamander (*Eurycea hillisi*), found in Alabama and Georgia; the western dwarf salamander (*Eurycea paludicola*), found in Louisiana, Texas, and Arkansas; and the bog dwarf salamander (*Eurycea sphagnicola*) of the Florida panhandle and coastal Alabama.

SIZE: Adults can reach 3½ inches long. Larval salamanders are generally under 1 inch.

Dwarf salamander.

DESCRIPTION: They have brown to yellow backs with dark sides. Salamanders in this species complex have four toes on the front two legs, distinguishing them from most other salamanders, which have five toes.

BEHAVIOR: Adults spend much of their time under leaves, moss, and woody debris on the edges of wetlands, particularly Carolina bays, pond edges, sandy streams, and swamp forests. Breeding occurs in these same wetlands, usually in winter, where larval salamanders hatch and spend several months before metamorphosing into miniature adults.

ADDITIONAL NOTES: The unique natural history and range information for each of these new species is still being gathered. Scientists sometimes refer to these visually similar but genetically distinct and closely related animals as a species complex.

Newts

Striped Newt (*Notophthalmus perstriatus*)
Broken-Striped Newt (*Notophthalmus viridescens dorsalis*)

Striped newts are found in longleaf pine forests in coastal and southern Georgia, central peninsular Florida, and a small area of Florida's eastern panhandle. This species is mostly restricted to temporary wetlands within fire-dependent communities, especially longleaf pine. Broken-striped newts are a subspecies of the wide-ranging red-spotted newt (*Notopthalmus viridescens*) that can be found across eastern North America. The broken-striped newt is mostly restricted to the southeastern coastal plain and can be especially abundant in longleaf pine forests.

SIZE: Both salamanders can reach about 4 inches long. Larval salamanders are generally under 1 inch. Eft-stage newts (the terrestrial juvenile stage) can range from 1 inch to nearly 4 inches.

DESCRIPTION: Adult striped newts are olive to brown, with red stripes running down the back. In the eft stage, it is generally bright red/orange, with stripes running down the dorsal surface. Adult broken-striped newts generally are yellow-olive-brown with a broken dorsal stripe running the length of the body. The eft stage is generally bright red/orange with a broken red stripe down the length of the body. Aquatic larvae of both species have large heads with external gills and similar patterning to the adult, although they are often duller in color.

BEHAVIOR: Adults of both species live in ponds and other wetlands. The eft stage is uncommon in the striped newt, and efts generally stay close to aquatic habitats. In the common eft stage, broken-striped newts may spend many years away from aquatic environments before returning to morph into breeding adults. Both species are toxic and lack many predators. Habitat loss is currently threatening striped newts. Little is known about striped newt breeding other than that the courtship can be lengthy and eggs are laid 1 at a time in aquatic vegetation. Broken-striped newts lay eggs in mass, with up to 300 eggs laid at one time. Broken-striped newts have elaborate courtship displays.

ADDITIONAL NOTES: These two species can be distinguished by the pattern on the dorsal surface. Striped newts have a solid line down the dorsal surface. Broken-striped newts have a line that is broken into smaller sections across the back.

Striped newt.

Broken-striped newt eft.

Broken-striped newt adult.

Gopher Frog

Lithobates capito

These frogs spend the better part of their lives underground in upland longleaf pine forests. They can be found in gopher tortoise burrows, the burrows of small mammals such as pocket gophers, and in stump holes. They require isolated, temporary wetlands for breeding sites with open canopy and emergent vegetation.

SIZE: This large, plump frog has a wide head. Adults are 2½–3½ inches long.

DESCRIPTION: The light-colored body is marked with dark brown or black blotches.

BEHAVIOR: Adults return to isolated wetlands only in the late fall, winter, and early spring to breed and lay eggs. Their call sounds like loud snoring. Their tadpoles develop in these wetlands and disperse back into longleaf uplands. Adult gopher frogs can sometimes live up to 1 mile away from breeding ponds and can spend their days underground, only emerging at night to feed.

ADDITIONAL NOTES: When exposed to bright light or when it feels threatened, this frog will put its front feet in front of its face to shield its eyes. The Mississippi gopher frog (*Lithobates sevosus*) is a visually similar species that is critically endangered, with only one wild population remaining in southern Mississippi.

Gopher frog.

Southern leopard frog.

Southern Leopard Frog

Lithobates sphenocephalus

This frog occurs throughout the longleaf pine range and extends well beyond into the mid-Atlantic coastal plain and southern Midwest. It is often found in pine flatwoods near or within small ponds.

SIZE: These midsized frogs reach a maximum size of about 3½ inches.

DESCRIPTION: This species is mostly gray to green, with large, randomly arranged circular dark spots on the back and legs.

BEHAVIOR: They breed in shallow wetlands such as ponds and brackish waters. Their mating call is a raspy gurgling croak. Females lay a large number of flattened eggs in shallow water. The tadpoles take more than 12 weeks to metamorphose from tadpole to frog.

ADDITIONAL NOTES: Southern leopard frogs most closely resemble pickerel frogs (*Lithobates palustris*), which overlap in part of the longleaf pine range. Instead of the southern leopard frogs' round spots down the back, pickerel frogs have square spots running in two distinct lines down the back. In addition, several new cryptic species of leopard frogs are being named, including the Atlantic coast leopard frog (*Lithobates kauffeldi*), that may extend into the range of longleaf pine.

Pine woods treefrog.

Pine Woods Treefrog
Hyla femoralis

A common inhabitant of the wet longleaf pine flatwoods, this frog can also be found in the more treeless pitcher plant bogs of the coastal plain. However, as it has a habit of staying high up in the tree, its distinctive "Morse code" or "riveting machine" call is often what makes it recognizable.

SIZE: This small-bodied frog is 1–1½ inches long.

DESCRIPTION: The body can be one of several colors, from gray to green to tan or brown, and may also be marked with irregular bands or blotches. Distinctive orange or light-yellow spots can be seen on the surfaces of the thighs.

BEHAVIOR: Like most amphibians, this frog needs shallow, fish-free wetlands for breeding. Enlarged, sticky toe pads on the feet allow this frog to easily climb to the top of longleaf pine trees. From their treetop perches, these frogs will ambush and feed on small arthropods.

ADDITIONAL NOTES: Sometimes these frogs can be seen perching on the lip of carnivorous pitcher plants, waiting for food.

Green Treefrog

Hyla cinerea

These moderately sized frogs live in southern swamps, marshes, other water bodies, and nearby uplands.

SIZE: Adults are 2½ inches long.

DESCRIPTION: They range from a dark gray-green to bright green, with a white-yellow stripe down the side of the body (lacking in some individuals) and with small yellow spots being present in some. Like most treefrogs, green treefrogs have long legs and sticky toe pads.

BEHAVIOR: Adults eat a variety of arthropods, occasionally feeding on small vertebrates such as other frogs. They breed throughout the summer in wetlands and can be found in large, noisy congregations. Often, they can be observed at outdoor lights in residential areas, feeding on the insects attracted to the lights. Green treefrogs are currently expanding their range.

ADDITIONAL NOTES: They look most similar to squirrel treefrogs (*Hyla squirella*). The two can be distinguished in most cases by the bright white-yellow stripe running down the green treefrog's side. The pine barrens treefrog (*Hyla andersonii*) might also be mistaken for a green treefrog or the squirrel treefrog.

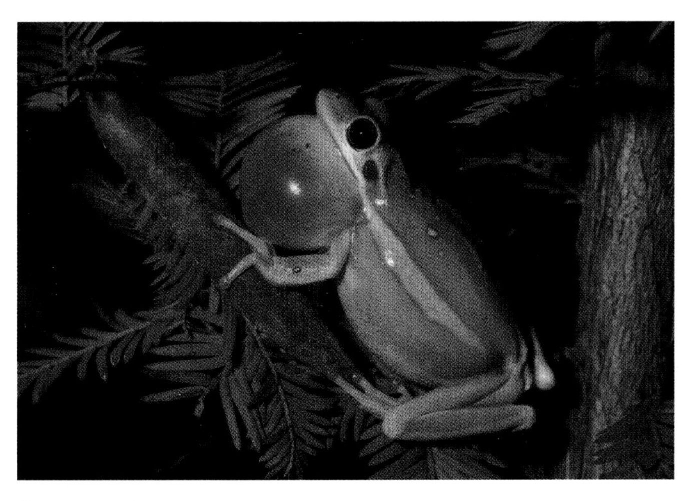

Green treefrog.

Pine barrens treefrog.

Barking Treefrog

Hyla gratiosa

These large treefrogs occur in scattered populations throughout most of the longleaf pine range. They are absent in the far west of the range and from most Piedmont communities.

SIZE: Barking treefrogs are the region's largest native treefrog species, reaching a maximum size of just over 3 inches with a heavy body.

DESCRIPTION: This treefrog ranges from gray to green, with conspicuous dark spots on the back and sides.

BEHAVIOR: Their distinct call from a distance sounds like barking dogs, especially when in mass. Breeding takes place in open wetlands where males float on the surface and call to attract a mate. Shallow wetlands such as Carolina bays or open fishless vernal pools are important breeding sites. Once breeding is over, adults spend most of their time high up in trees and can be found far into the uplands away from their breeding sites.

ADDITIONAL NOTES: Squirrel and green treefrogs are most similar in coloration but lack the abundant large spots seen on the back of barking treefrogs. In addition, barking treefrogs are more heavily bodied and have a squarish snout not present in the other two species. In some portions of their range, barking treefrogs have declined due to habitat loss.

Barking treefrog.

Squirrel treefrog.

Squirrel Treefrog
Hyla squirella

These small treefrogs occur throughout most of longleaf pine's range and can be especially abundant in the dense wetland stands associated with longleaf pine forests.

SIZE: They are small, reaching a maximum size of 1½ inches.

DESCRIPTION: Squirrel treefrogs can be highly variable in appearance, often requiring a process of elimination to identify them. The color can range from gray to green but also can appear golden or bronze under certain conditions. Some individuals may have a bright yellow patch on the back of the legs. Some individuals are plain with no spots, while others can be heavily patterned with small spots across the back and legs.

BEHAVIOR: These treefrogs get their name from a raspy call resembling that of a squirrel. In some areas these frogs can be incredibly abundant, especially in urban environments. In longleaf pine forests, they often congregate around fishless wetlands with dense vegetation. During breeding, many males may congregate and compete for mates. Breeding takes place anytime between March and October. Males may also call during heavy rains.

ADDITIONAL NOTES: As these treefrogs are highly variable in appearance, they can be confused with many other species. They lack the white stripe down the side seen in green treefrogs. They lack the numerous large dark spots seen in barking treefrogs and are smaller bodied, with a more rounded snout. They usually lack the extensive mottling seen in pine woods treefrogs and brilliant spots on the underside of the hind legs, although they may also have some yellow coloration.

Southern chorus frog.

Southern Chorus Frog

Pseudacris nigrita

These small frogs extend into the coastal plain from southern North Carolina to far eastern Louisiana.

SIZE: Their maximum size is about 1 inch.

DESCRIPTION: This frog is mostly gray, with some mottling on the back and sides, and often with darker coloration on the lower portions of the sides. The ventral surface is generally white to cream.

BEHAVIOR: They are often the first sign of spring in longleaf pine forests, calling on warm winter days. Adults gather in shallow wetlands, pools, and ditches, where males call to attract a mate. Females deposit eggs on vegetation or debris in small clusters; these are then externally fertilized by the males. Tadpoles generally undergo metamorphosis within 2 months following fertilization.

ADDITIONAL NOTES: Southern chorus frogs are the most widespread of the chorus frogs within the longleaf pine range, but Brimley's chorus frog (*Pseudacris brimleyi*) also occurs in longleaf pine forests in the coastal plains of Virginia, North Carolina, South Carolina, and Georgia. Brimley's chorus frog is distinguished by the dark stripe running down each side from the snout to the belly. The range of ornate chorus frogs (*Pseudacris ornata*) is very similar to the southern chorus frog, but it can be identified by the ornate, almost black blotches on the sides and around the eye.

Little Grass Frog

Pseudacris ocularis

Little grass frog.

Little grass frogs are small frogs of grassy wetlands occurring in the coastal plains of southern Virginia, North Carolina, South Carolina, Georgia, northern Florida, and far southeastern Alabama.

SIZE: Their maximum size is ½ inch, making them the smallest frog native to North America.

DESCRIPTION: This species is generally tan-brown, with a dark stripe running from each eye down the sides.

BEHAVIOR: These frogs may breed throughout the year, but most breeding takes place in spring. Calls may be heard year-round even when no breeding is taking place. Eggs are laid in shallow, grassy wetlands. Tadpoles metamorphose to frogs in as little as 6 weeks.

ADDITIONAL NOTES: Their diminutive size and dark stripe are diagnostic, with perhaps only Brimley's chorus frog having similar coloration, but Brimley's is generally more than double the size of the little grass frog.

Spring Peeper

Pseudacris crucifer

Spring peeper.

Spring peepers occur throughout almost all of the eastern United States and within the longleaf pine range. They are often heard at small ponds and wetlands.

SIZE: These small frogs reach a maximum size of approximately 1 inch.

DESCRIPTION: This small species is generally a tan to orangish color with a distinct X pattern on the back.

BEHAVIOR: These frogs breed in wetlands in late winter to early spring. Their call is very loud in relation to their small size and is a recognizable sound of spring throughout much of the eastern United States. Breeding takes place on the margins of wetlands, streams, ponds, and swamp forests. Males call to attract a female who attaches her eggs to floating vegetation or debris. After breeding, spring peepers move into surrounding uplands.

ADDITIONAL NOTES: Old publications like *The Old Farmer's Almanac* indicate that the "peep, peep, peep" call of this chorus frog is one the first signs of spring. However, in many areas of the longleaf pine range that do not experience prolonged freezes, spring peepers can be heard as early as late November.

Southern Toad

Anaxyrus terrestris

These robust amphibians with their warty skin are common throughout much of the Southeast. They can be found in a variety of habitats but are well suited for dry longleaf pine forests, where other amphibians may not survive.

SIZE: They generally reach about 3 inches in length, but individuals over 4 inches have been recorded.

DESCRIPTION: Their highly variable coloration ranges from charcoal to light gray and from brown to rust red. Most individuals also have some spotting on the back.

BEHAVIOR: Breeding takes place in spring, with many individuals congregating at a site with shallow, standing water. Thousands of eggs can be laid by a single individual; these hatch in only a few days and metamorphose from

Southern toad.

tadpole to toad in 1–3 months. The call is a bright trill similar to many other toad species, though higher in pitch than the American toad.

ADDITIONAL NOTES: Southern toads have a cranial crest that runs between the parotid gland (kidney-shaped gland behind the eye and ear). While present in the American toad (*Anaxyrus americanus*), this cranial crest is much more pronounced in the southern toad and distinguishes the two species.

Oak Toad

Anaxyrus quercicus

These are small toads of longleaf pine and associated habitats through most of the longleaf pine range. They are absent from the far western portion and from most of the Piedmont region.

SIZE: This is the smallest toad in North America, no more than about 1¾ inches.

DESCRIPTION: They are generally gray-charcoal, with some red speckling and a distinct mid-dorsal stripe just over the spinal column. They have large parotid glands (kidney-shaped glands behind the eye) relative to their size.

BEHAVIOR: Oak toads are more active during the day than other southeastern toads. They breed in shallow pools in pine forests and associated

Oak toad.

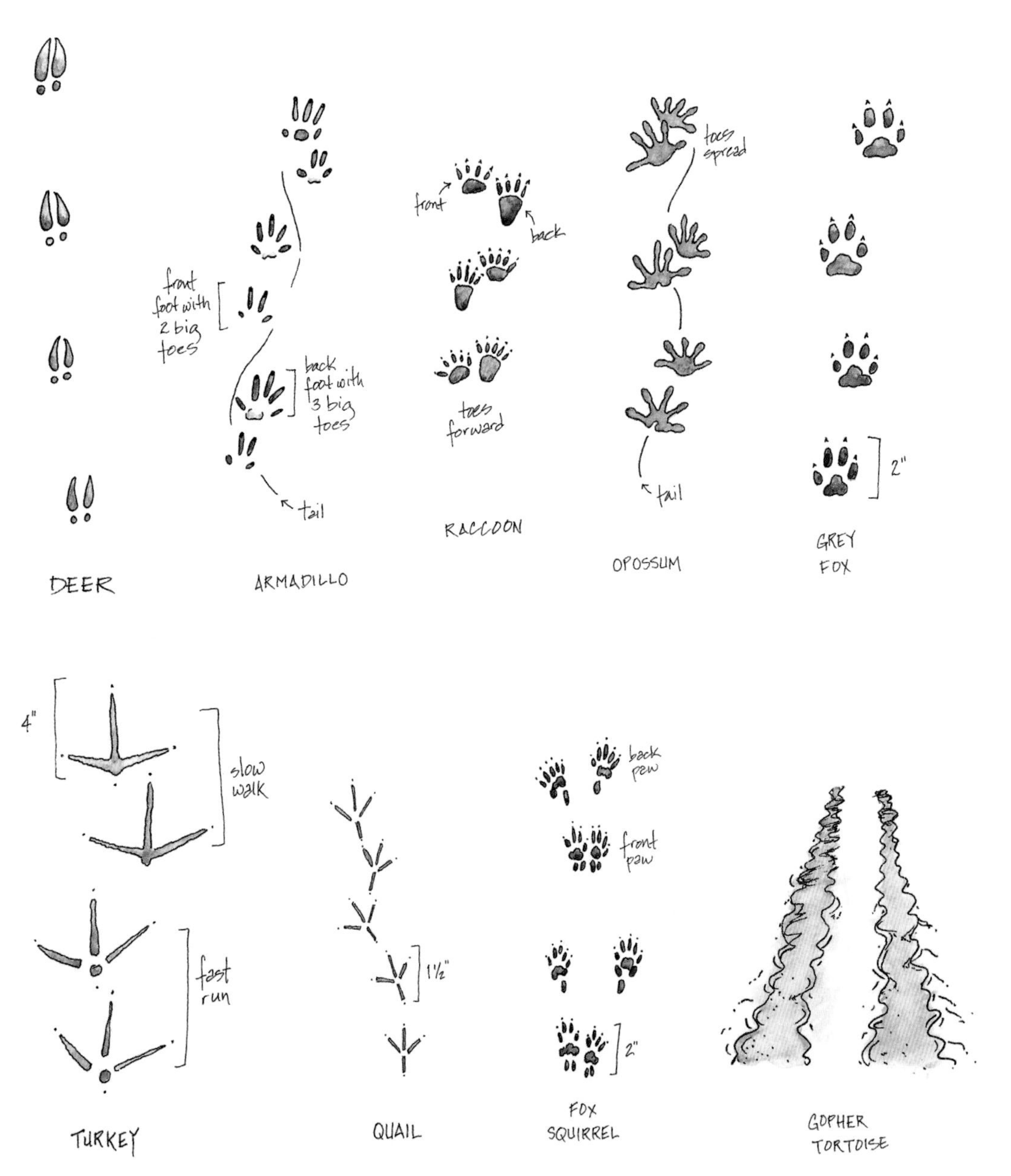

The tracks on sandy roads or fresh firebreaks can sometimes reveal what animals are living in a forest. Look for the footprints of these common wildlife species found in longleaf pine forests.

communities such as oak scrub. Males produce a high-pitched chirping call to attract potential mates; some describe the sound as similar to that made by a "baby chick." Females lay small strands of eggs that are either attached to vegetation or free floating and are fertilized externally by males. Tadpoles can metamorphose to toads in as little as 4 weeks.

ADDITIONAL NOTES: Scientists are still trying to determine why these toads have declined in recent years. This species was once abundant in the coastal plain, but in some areas it has been completely eradicated.

Eastern Spadefoot

Scaphiopus holbrookii

These secretive amphibians occur throughout most of the longleaf pine range. They are robust and highly adaptable.

SIZE: Moderate in size, they may reach up to 3½ inches.

DESCRIPTION: Adults have moist skin with a warty texture. They are generally gray, brown, or yellowish but may have reddish spots. In most individuals, especially males, two lines run from the eyes down the dorsal surface, forming an hourglass shape. In addition, the yellow stripes running down the back in most individuals is diagnostic. Their brilliant golden eyes with catlike pupils are one distinguishing characteristic from true toads. The name comes from sickle-shaped hardened projections on the back feet that help them dig.

Eastern spadefoot.

BEHAVIOR: Adult spadefoots spend a large amount of time underground in their burrows. They come out on rainy, moist evenings to feed or mate. Mating occurs in mass, with many adults congregating in each breeding site. Males float on the surface, calling to attract a mate. Females may deposit thousands of eggs at once; these are then fertilized by a male. These spadefoots are thought to live long lives that allow them to persist even in urban and agricultural areas. The tadpoles have the shortest metamorphosis time of any frog in the United States. Tadpoles in stable conditions will be mostly vegetarian. If the wetland begins to dry, they will shift to a cannibalistic form to obtain enough energy to metamorphose.

ADDITIONAL NOTES: Few species resemble eastern spadefoot within their range. Their eyes have catlike pupils to help them see in the dark. Toxic skin secretions can cause mild reactions in people and pets.

Eastern Narrowmouth Toad

Gastrophryne carolinensis

These small toads with a flattened body and pointed snout occur throughout the entire longleaf pine range. They can occupy both dry and damp habitats.

Eastern narrowmouth toad.

SIZE: Narrowmouth toads are small, no more than 2 inches.

DESCRIPTION: Adults have moist skin with a flattened body, a fold of skin just behind the head, and small round eyes. They are usually gray-brown to bronze on their dorsal surface, with light speckling. They generally have darker coloration on their sides and ventral surface.

BEHAVIOR: Adults are somewhat secretive, mostly occurring underneath cover, often seeking moisture even in dry habitats. They are specialists that feed on a variety of small ants and occasionally other invertebrates. Breeding rituals take place on rainy evenings in shallow pools. The male's loud call can be heard far away and is often referred to as a nasal "waaaaaah-hhhhhhhhhhhhhh." This sound is sometimes compared to a sheep, giving this toad the colloquial name of sheep frog. The eggs float on the water surface of the pool and can take as little as 20 days to metamorphose from tadpoles to adult.

ADDITIONAL NOTES: The sharp, pointed nose is distinctive compared to the region's other frogs and toads. The call is often the best indication that this species is present in an area.

Invertebrates

Butterflies

The fire-maintained, open-canopied longleaf forest is typically filled with the wildflowers of many plant species, especially during the summer and fall seasons. This meadow-like scene is enhanced by the presence of scores of "flying flowers." Colorful butterflies flit from flower to flower, feeding and pollinating in the process. Many of these species are attracted to the flowers of specific nectar plant species that act as energy sources for the butterflies. The larvae, or caterpillars, feed voraciously on the leaves and stems of a variety of host plants. Some caterpillars are host specific, feeding only on one particular type of host plant.

1. Butterflies with plant associates (energy sources and host plants). Clockwise from top right: monarch butterfly (*Danaus plexippus*) with butterfly weed (*Asclepias tuberosa*); queen butterfly (*Danaus gilippus*) with swamp milkweed (*Asclepias incarnata*) and desmodium (*Desmodium* spp.); red-spotted purple butterfly (*Limenitis arthemis*) with deerberry (*Vaccinium stamineum*) and dung; spicebush swallowtail butterfly (*Pterourus troilus*) with sassafras (*Sassafras albidum*) and azalea (*Azalea* spp.); zebra swallowtail butterfly (*Euryotides marcellus*) with pawpaw (*Asimina triloba*) and azalea (*Rhododendron* spp.); tiger swallowtail (*Papilio glaucus*) with southern magnolia (*Magnolia grandiflora*) and blazing star (*Liatris* spp.).

2. Butterflies with plant associates (energy sources and host plants). Clockwise from top right: Gulf fritillary butterfly (*Agraulis vanillae*) with passionflower (*Passiflora incarnata*) and goldenrod (*Solidago* spp.); eastern tailed blue butterfly (*Cupido comyntas*) with goat's rue (*Tephrosia virginiana*) and aster (*Symphyotrichum* spp.); zebra longwing butterfly (*Heliconius charithonia*) with passionflower and goldenrod; cloudless sulfur butterfly (*Phoebis sennae*) with partridge pea (*Chamaecrista fasciculata*) and scarlet sage (*Salvia coccinea*); little metalmark butterfly (*Calephelis virginiensis*) with vanillaleaf (*Carphephorus odoratissimus*) and aster family plants; common buckeye (*Junonia coenia*) with foxglove (*Penstemon* spp.) and boneset (*Eupatorium* spp.).

Velvet Ant

Dasymutilla occidentalis

This wasp species is a common resident in longleaf pine forests, especially in areas of deep sand. Females lack wings and resemble ants. It is this similarity of appearance that gives them the common name "velvet ant." People most commonly encounter females; the winged males are rarely noticed.

SIZE: Adults reach about ¾ inch in length.

COLOR: Adults have black exoskeletons with dense red-orange hairs covering the head–thorax. Males have dark-colored wings.

BEHAVIOR: Adults primarily feed on nectar. Females lay eggs in the larvae or cocoons of other insects such as wasps, beetles, or flies. When hatched, the velvet ant larva eats its host. In addition to warning coloration, and the female's painful sting, both males and females emit a squeaking noise as a defense strategy when threatened.

ADDITIONAL NOTES: The female's sting is so painful that people once thought it was potent enough to kill cattle, giving them another common name, "cow killer." However, these wasps do not possess venom toxic enough to kill large mammals.

Female velvet ant.

Many different kinds of pollinators can be found in longleaf pine forests, including bees, wasps, hummingbirds, and butterflies. Perhaps the most well known of the pollinators, bees, can be found in longleaf pine forests in great diversity. Bees are pollinators not only for many of the native wildflowers but also for many of the surrounding crop and agricultural lands. Hundreds of different bee species are found in these forests, representing many different families, such as sweat bees (*Halictidae*) and true bees (*Apidae*). The nonnative European honeybee (producing the kind of honey consumed by humans) is uncommon in longleaf pine forests unless an apiary is nearby.

Golden green sweat bee.

Sweat bees are often small, range in color from metallic green to teal, and nest underground. They don't produce honey but may sting if provoked. Some of the more common sweat bees include the tiny golden green sweat bee (*Augochlorella aurata*), reticulate metallic sweat bee (*Lasioglossum reticulatum*), and the rugose-chested sweat bee (*Lasioglossum pectorale*). They can be found on native flowers such as asters and milkweeds. They are often overlooked until they land on your sweaty arm in the warmer months. However, they are the most common family of bees in longleaf woods.

Common long-horned bee.

The second-most common family of bees is diverse in size and classification. These are the bees in *Apidae*, or true bees, which includes carpenter bees, digger bees, and bumblebees. Two of the more common bees of this family are the Florida small carpenter bee (*Certaina floridana*) and the common long-horned bee (*Melissodes communis*). They are frequently found feeding on plants like goldenrods and nesting in shrubs like sumac. Digger bees are small solitary ground-dwelling bees that use open soil to build nests. In some cases, they may form extensive colonies of solitary nests. Bumblebees range in size from the small and abundant common eastern bumblebee (*Bombus impatiens*) to the large and threatened southern plains bumblebee (*Bombus fraternus*). Several bumblebee species have declined sharply in the past 50 years and now face extreme conservation challenges. The decline of longleaf pine forests may be a contributing factor for some species.

Southern plains bumblebee.

Invertebrate Commensals of Gopher Tortoise Burrows

Animals that obtain food, refuge, and other benefits from the burrows of gopher tortoises are known as commensals. Over 300 species of commensals have been documented to use the burrows of gopher tortoises. Vertebrate animals like toads, snakes, mice, moles, and even birds are more often observed mostly because of their size. Invertebrate species that use these burrows, however, are often less well known and very rarely photographed. There are several species of beetles, flies, moths, and even crickets unique to these burrows.

Several species of scarab beetles occur in longleaf pine forests with the potential for new discoveries. Some feed on dead animals. Others are dung beetles that are highly specialized, feeding solely on the feces of a specific species of animal. Several particularly specialized species rely on gopher tortoise burrows and dung for part or all of their life cycles. At least four species of scarab beetles live solely in gopher tortoise burrows and feed exclusively on gopher tortoise dung: gopher tortoise copris beetle (*Copris gopher*), gopher tortoise onthophagus beetle (*Onthophagus polyphemi*), the little gopher tortoise scarab beetle (*Alloblackburneus troglodytes*), and the gopher tortoise aphodius beetle (*Aphodius laevigatus*). The gopher tortoise hister beetle (*Chelyoxenus xerobatis*) also uses the dung of gopher tortoises, though it is not a scarab beetle.

Camel crickets are functionally wingless insects that are sometimes associated with gopher tortoise burrows. Two of the better known camel crickets to use these burrows are gopher crickets (*Ceuthophilus latibuli*) and Walker's camel cricket (*Ceuthophilus walkeri*). These crickets can provide a food source for other animals using tortoise burrows such as gopher frogs.

All of these invertebrates described can sometimes be seen if you have a camera that goes down into a gopher tortoise burrow. Otherwise, you won't see them moving out and about in longleaf pine forests.

A sampling of the many insect and animal species that use gopher tortoise burrows for habitat.

Gopher Tick

Amblyomma tuberculatum

Gopher tick.

As the name implies, these ticks can be found on gopher tortoises. They are commonly found latched onto the gopher tortoise on the soft skin parts not protected by its shell. This tick uses the gopher tortoise as its primary host. As with the tortoise, the gopher tick numbers are thought to have declined.

SIZE: Typically, these ticks are ¼ inch long but can be up to 1 inch when engorged with blood.

COLOR: The body is buff, which is more notable when the tick is engorged with blood. Head and legs are rust colored.

BEHAVIOR: It is typically found near the entrance of gopher tortoise burrows during the warm summer months. Like most ticks, these species latch onto their host and siphon off blood. Female gopher ticks have been recorded to lay close to 10,000 eggs at once.

ADDITIONAL NOTES: Several species of ticks can be found moving about longleaf pine forests or other forests and grasslands of the South. These include the lone star tick (*Amblyomma americanum*), black-legged tick or deer tick (*Ixodes scapularis*), brown dog tick (*Rhipicephalus sanguineus*), and Gulf Coast tick (*Amblyomma maculatum*). These species will readily attach to humans. Some can transfer diseases, including Lyme disease and Rocky Mountain spotted fever. It's uncertain if gopher ticks carry such diseases.

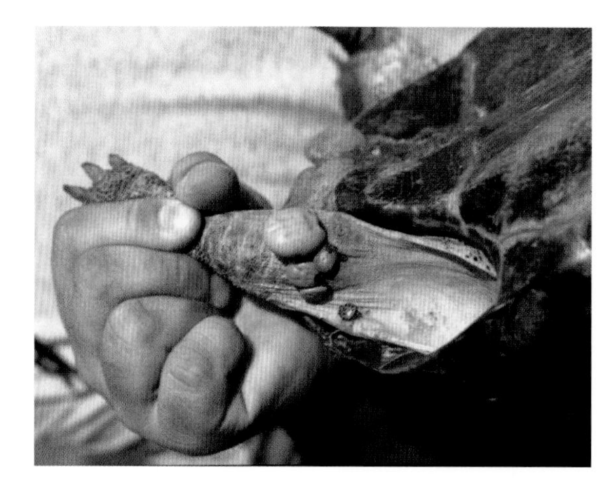

Gopher ticks attached and engorged on the rear leg of a gopher tortoise.

Black Turpentine Beetle
Dendroctonus terebrans

Black turpentine beetle.

Though healthy longleaf pine trees are extraordinarily resistant to the pine beetle, trees under stress from lightning strikes, wind damage, and drought are at higher risk of being infested by black turpentine beetles. Healthy longleaf pine trees produce ample resin that pushes beetles out of the tree. Beetle presence in a longleaf pine is recognized by their pitch tubes on the lower 8 feet of the tree. Tubes are large and look similar to a shelled walnut growing from the side of the tree. A large presence of these beetles in a longleaf pine tree often means it will die soon.

SIZE: Adults are ¼–⅓ inch in length. Larvae are ¼ inch.

COLOR: Adult beetles are dark brown or black, while larvae are creamy white with a reddish-brown head.

BEHAVIOR: Adults bore entrance holes in the bark of the pine tree and chew out tunnels between the outer bark and wood of the tree. Eggs are laid in these tunnels, and larvae hatch within 10 days. The larvae feed for several weeks and eventually mature into adult beetles. The new adults emerge from holes chewed through the bark. This life cycle is typically 3–4 months.

ADDITIONAL NOTES: Woodpeckers are not thought to feed heavily on black turpentine beetles, unlike other beetles found on longleaf pine trees. That said, if black turpentine beetles are found on a longleaf pine tree, southern pine beetles (*Dendroctonus frontalis*), small southern pine engraver (*Ips avulsus*), eastern five-spined engraver (*Ips grandicollis*), and coarse-writing engraver (*Ips calligraphus*) will usually be found elsewhere on the tree, often higher up.

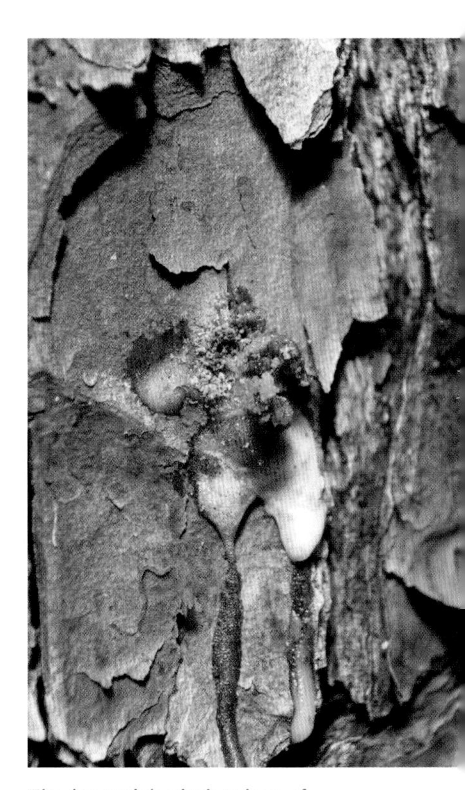
The brownish pitch tubes of the black turpentine beetle are usually found near the base of the tree.

Sculpted Pine Borer
Chalcophora virginiensis

Like the bark beetles, the group of insects known as metallic wood borers are common residents of longleaf pine forests but cause little harm to healthy longleaf pine trees. Of all the species in the South, the sculpted pine borer is among the most common. They can be found where there are many dead or dying pines, as they are attracted to the scent of the natural terpenes (think of Pine-sol®) emitted by pines in distress.

Sculpted pine borer.

SIZE: They are up to 1½ inches in length with a flattened body.

COLOR: Adults have a black-to-copper metallic luster. Larvae are cream colored.

BEHAVIOR: The larvae of this wood-boring species tunnel beneath bark and into the wood of the tree. They produce oval or flattened tunnels as they chew their way through the wood. These larvae can be an important food source to a variety of woodpeckers found in longleaf pine forests.

ADDITIONAL NOTES: The term "wood borer" refers to this insect when it is in the larval form. The larvae can bore into the wood of dead longleaf pines. The sculpted pine borer is also known as a flathead borer. Another wood borer in longleaf pine trees is the southern pine sawyer (*Monochamus titillator*). The larvae of these long-horned beetles are considered round-headed borers and can be prevalent following hurricanes. Another wood-boring insect, the turpentine borer (*Buprestis apricans*), was once problematic during the naval stores industry times when scraping off excess gum stressed the trees.

Florida Harvester Ants
Pogonomyrmex badius

Though several species of ants may be observed in the sandy soils of longleaf pine forests, Florida harvester ants are easily identified by their unique mounds. These are fairly flat, with minimal entrances in the center. The area

Florida harvester ants. The major worker ant (right) has an enlarged muscular head for milling seeds. Soldier ants (left) are smaller.

Florida harvester ant mound. Near burned areas, the edge of the mounds are often covered with small pebbles or charcoal.

around the mound is typically kept clear of vegetation yet will be covered with small pebbles or charcoal from burned areas.

SIZE: They vary from ¼ to ⅓ inch long.

COLOR: The adults are a dark rusty red.

BEHAVIOR: Specialized worker ants known as soldier ants venture into the woods to collect fallen seeds or even pluck seeds directly from plants. Longleaf pine seeds are not part of the harvester ants' diet. They instead prefer smaller-seeded grasses and sedges. Seeds are cleaned, and leftover plant fiber is deposited at the edge of the mound.

ADDITIONAL NOTES: Some have speculated that baits and chemicals used to control invasive, nonnative fire ants have caused harvester ant populations to also significantly decline in areas.

Cone Ants

Dorymyrmex bureni

This is one of the most common native ants found in the sandy soils where longleaf pines grow. Compared with imported fire ants found in similar areas, their nests are organized into well-shaped cones. Their nests look like miniature volcanoes, with a crater-shaped mound and single entrance hole. The fast-moving cone ant is not aggressive toward humans. Imported fire ants, by contrast, are exceptionally aggressive and need little provocation to attack humans.

SIZE: The body length is up to ⅛ inch.

COLOR: This ant is light orange.

BEHAVIOR: These ants eat live insects, including other ants. They are attracted to sugars and can oftentimes be found close to the honeydew (i.e., insect poo) of aphids. It is not uncommon to see them on milkweed plants collecting honeydew from their livestock-like aphids.

ADDITIONAL NOTES: When crushed, these ants emit a foul odor, earning them the nickname of piss ant in many areas of the South.

Cone ant workers ants carrying sand.

The distinctive conical nest of the cone ant.

Ants: The Unseen Denizens

Ants are often one of the most overlooked insects in longleaf pine forests unless you're bitten or stung by one. By and large, ants are often hard to distinguish except by experts, which adds to their obscurity. However, all native ants serve vital functions in the forest: soil aeration, seed dispersal, and as a food source. Ant diversity in longleaf pine forests is among the highest reported in North America. Despite that, scientists are still finding new species, such as the myrmicine ant (*Temnothorax palustris*) in Florida's longleaf pine flatwoods.

Small environmental changes can determine the niche occupied by any particular ant species. For example, grass- versus shrub-dominated groundcover, soil type, or even size (or species) of trees can all determine which ant species is present. Some ants live solely in trees (arboreal), forage just on the ground, or are subterranean (underground). The acrobat ant (*Crematogaster ashmeadi*) is one of the more abundant ants in longleaf pine forests and is an important food source of the Red-cockaded Woodpecker (*Picoides borealis*). Like the acrobat ant, the carpenter ant (*Camponotus nearcticus*) is also considered an arboreal ant. Contrast that with ant species like the vampire ant (*Stigmatomma pallipes*) or thief ant (*Solenopsis carolinensis*) that are completely subterranean. There are also many species of ants that are neither subterranean nor arboreal but are primarily foraging ants, including *Pheidole morrisi* (one of most abundant ants), cone ants (*Dorymyrmex bureni*), high noon ant (*Forelius pruinosus*), and field ant (*Formica pallidefulva*).

Carpenter ant worker standing above the inconspicuous entrance hole to its colony.

The effect of nonnative ants on native ant diversity is not well known. However, some nonnative ants, such as the red imported fire ant (*Solenopsis invicta*), black imported fire ant (*Solenopsis richteri*), and the hybrid of the two species, have been shown to dominate some areas of longleaf pine forest. Other nonnative species, such as the tawny crazy ant (*Nylanderia fulva*), are just beginning to occupy areas of longleaf pine forest. These aggressive, nonnative ants could have a negative long-term effect on the diversity of the longleaf pine forest.

Green Lynx Spider
Peucetia viridans

Green lynx spider.

Wet pine savannas may contain unique carnivorous plant bogs, including a variety of *Sarracenia* (pitcher plant) species. These fluted plants can be a death sentence for many flying insects and ants. However, the green lynx spider has adapted to use these pitcher plants as hunting grounds. This spider can frequently be found near the flute of the pitcher plants.

SIZE: Females can be up to ⅝ inches and males up to ½ inches long. Males are often slenderer.

COLOR: The body is a bright green that can often match the green plants the spider is on. Legs can have either a yellow or pale green with black spines. On the abdomen's dorsal surface, many chevron-like marks can be seen.

BEHAVIOR: The green lynx spider does not spin a web like many spiders but instead relies on its camouflaging color, speed, and agility to be a skilled ambush predator. This spider possesses venom that is relatively harmless to humans. The female spiders have been reported to spit this venom in self-defense.

ADDITIONAL NOTES: These spiders can be found lying in ambush near the mouth of a pitcher plant. Though they are not known to have an obligate relationship, the spiders are opportunistic robbers waiting for insect prey to be attracted to the pitcher plant.

Golden Orb Weaver Spider
Trichonephila clavipes

This large spider species is well known to many hunters and outdoors people. Its large golden web can often be spun at head level for humans. If you've spent much time in the woods, you have inevitably walked face first through its web.

SIZE: Females are noticeably larger than males. With legs included, the size of some larger females can be over 5 inches. Males are significantly smaller and can sometimes be seen on the webs of female spiders.

COLOR: The female spider is conspicuous in not only size but color. The abdomen is yellow/brown with white spots. Legs are also yellow with brown hairy bands. The head is silver. Males are an unassuming dark brown.

BEHAVIOR: In the late summer and fall, these spiders construct webs 3–6 feet in diameter. These large, sticky webs trap a variety of medium-sized flying insects, including small beetles, dragonflies, flies, bees, and wasps. The spider is mildly venomous to humans but will often only bite when held or pinched.

ADDITIONAL NOTES: The genus, *Trichonephila*, means "fond of spinning." They may also be known as golden silk spiders or giant wood spiders.

Female golden orb weaver spider.

Spiny Backed Orb Weaver

Gasteracantha cancriformis

These colorful spiders may be especially common during the fall and winter months over part of their range. They weave an intricate web that is easily visible with morning dew on it. Though these spiders can look dangerous with their spines and bright colors, they are, in fact, harmless to humans. Males and females differ distinctly in size, shape, and color.

SIZE: Males measure under 1⁄10 inch in length and are about as wide as they are long. Females may be up to ½ inch long and ⅓ inch wide.

COLOR: Females have bright coloration relative to males, but geographic variation in color does occur. The back (dorsum) of females is mostly white but can also be yellow and sometimes red. Black spots predominate. Individuals have six pointed spines that are usually red, but black ones can be found as well. The back is white with black spots. Males lack spines, and the back is gray or black with white spots.

BEHAVIOR: The female weaves the web to capture flying prey such as horseflies, beetles, mosquitoes, and moths. Some have speculated that tufts of silk are woven into the web to make the web easier to see so that larger animals (birds, bats, humans, etc.) will avoid it. The male waits in the female's web for an opportunity to breed. As with many spider species, the male may become dinner when mating is over.

ADDITIONAL NOTES: Though they can be called crab spiders locally, they are not true crab spiders.

Spiny backed orb weaver.

Giant wolf spider inside a gopher tortoise burrow.

Giant Wolf Spider

Hogna carolinensis

These are common ground spiders in longleaf pine forests. The term "wolf spider" covers a multitude of species that share common traits. These spiders do not construct large, suspended webs but instead are seen on the ground, in burrows, or even on the side of trees. The more common Carolina wolf spider is the largest of the wolf spiders in the Southeast.

SIZE: This is a large spider, measuring 1–1½ inches in length.

COLOR: These spiders are well camouflaged but typically are a combination of brown, gray, and black.

BEHAVIOR: This spider uses its eight eyes, arranged in three rows, for hunting. Unlike other spiders that use their webs to trap prey, the Carolina wolf spider relies on good eyesight to find a meal. They hunt alone.

ADDITIONAL NOTES: This is the state spider of South Carolina.

American Grass Mantis

Thesprotia graminis

Though this is a mantid, it is often mistaken for a stick insect. However, when it folds its front legs for ambush, there is no question that it belongs in the praying mantis family, albeit a very skinny cousin with a tiny head and legs. Their nondescript coloration provides camouflage for hunting and serves to hide them from predators. The wingless female American grass mantis bears an uncanny resemblance to the longleaf pine needle. As such, they are not always the easiest insect to see in the woods.

SIZE: Males grow to 2 inches long, and the larger females can be 2½–3 inches.

COLOR: It is usually light brown.

BEHAVIOR: Males have well-formed wings and are quite adept at flight. The larger females, however, can't fly. That does not stop either sex from being efficient and deadly predators. They ambush prey including moths, crickets, beetles, and spiders.

ADDITIONAL NOTES: Earning a bad reputation in South Carolina folklore, the mantis is considered an agent of the devil. If it spits in your eye, you'll go blind, or so the folklore goes.

American grass mantis.

American grass mantis blending into pine straw.

Southern Two-Striped Walking Stick

Anisomorpha buprestoides

These phasmids are large insects with variable coloration across their range. This species can be distinguished from other phasmids by two prominent stripes running down the body. The insect can sometimes be quite abundant, especially during mating season. Outside mating season, they are often found under loose bark of both living and dead longleaf pine trees.

SIZE: Females are comparatively larger than males, growing up to 4 inches long while males are rarely found longer than 1½ inches.

COLOR: Three distinct colors can be found throughout its range in the Southeast: brown, white, and orange. Different color forms become more prevalent in certain types of ecosystems. The brown form is especially common in longleaf pine forests.

BEHAVIOR: The mating behavior of this insect is unique: the much smaller male couples with the female and hitches a ride on her back as she goes about her daily routine. The southern two-striped walking stick can frequently be found on oak trees such as turkey and red oak. It can also be found eating blackberry bushes.

The smaller male and female southern two-striped walking stick.

ADDITIONAL NOTES: When threatened, this insect can squirt a strong-smelling, potent, defensive spray. It has been documented that, if sprayed in the eyes of the attacker (be it dog or human or other), the ill effects can last for several days. They are also known by the names of devil rider or palmetto walking stick.

Florida Blue Centipede

Scolopendra viridis

It's unlikely to see this large centipede moving about during the day. Most daytime observations occur by lifting rocks or partially rotten logs. Though they can be fierce predators, centipedes also serve as a vital food source for other animals, including the Red-cockaded Woodpecker.

SIZE: This centipede can measure up to 3 inches in length.

COLOR: As the name implies, the Florida blue centipede is bluish gray.

BEHAVIOR: This predatory insect catches prey with its jaws and then kills it by injecting venom. It will feed on earthworms, slugs, snails, crickets, spiders, millipedes, and other arthropods. The female may provide some parental care to eggs before they hatch.

ADDITIONAL NOTES: These centipedes have been known to bite humans. Their bite would be similar to that of a bee sting, with localized swelling, itchiness, and redness.

Florida blue centipede.

American Grasshopper

Schistocerca americana

Grasshoppers are a common sight in longleaf pine woodlands. They are especially noticeable flying ahead of fire moving through the grasses. Among the larger and more common grasshoppers is the American grasshopper. In many areas of the longleaf pine range, this insect can be found year-round.

SIZE: Males are 1½–1¾ inches long. Females are larger, from 1¾ inches to slightly over 2 inches.

COLOR: The adult is yellowish brown. Wings are paler, with large brown spots. A cream-colored stripe extends the length of the body. Nymphs (young

American grasshopper.

insects) can vary from green to yellow to red. Color is said to be determined largely by temperature differences.

BEHAVIOR: These grasshoppers feed primarily on leaves, often eating holes in them. Their densities can become large enough to become a pest. Adults have fully developed wings and are adept at flying. Due to their size (and possibly taste), only a few of the larger birds and more aggressive birds (such as Northern Mockingbirds) have been documented eating them.

ADDITIONAL NOTES: This grasshopper is known to fly up into trees when disturbed, earning it another common name of bird grasshopper.

Eastern Lubber Grasshopper

Romalea guttata

These large, flightless grasshoppers use walking as the primary form of locomotion. Like many grasshoppers, the eastern lubber can cause localized damage to plants in longleaf pine forests. The mouth parts are more adapted to eating broadleaf plants rather than grasses.

SIZE: Males are 1⅔–2 inches long; females are 2–2¾ inches.

COLOR: Color variation occurs across its range. Some lubbers are yellow with black accents, and some are black with yellow accents. Legs are reddish brown. The spindly wings (too small for flight) are rose to pink.

Eastern lubber grasshopper.

BEHAVIOR: These grasshoppers are toxic to most potential predators. When the lubber is alarmed, it will hiss, secrete a foul-smelling and toxic froth, and spread its brightly colored wings.

ADDITIONAL NOTES: Loggerhead Shrikes have been observed impaling these grasshoppers on fences and thorns and leaving their carcass there for a few days. This curing time is thought to allow the toxins in the grasshoppers to dissipate and become digestible to the shrike.

Roaches

Most of us aren't enthusiastic to see roaches in our houses. However, in longleaf pine forests, these insects are a very important part of the food chain. Many animals, like the Red-cockaded Woodpecker, readily eat roaches. Two roaches of particular abundance are the fulvous wood roach and the Florida woods cockroach (also called a palmetto bug or skunk cockroach). The animals that eat them are often efficient at finding them under places like loose bark, decaying logs, leaves, and straw. The Florida woods cockroach has a pungent defensive secretion, earning it the nickname skunk roach. The males of both species have wings for flight.

Fulvous wood roach.

Florida woods cockroach.

Fulvous Wood Roach (*Parcoblatta fulvescens*)

SIZE: On average, this species measures ½–⅔ inch long, with males being slenderer and slightly longer than females.

COLOR: Males are a uniform pale brown/yellow. Females are a gradient of orange, red, and black.

Florida Woods Cockroach (*Eurycotis floridana*)

SIZE: These are large bodied, averaging 1¼–1⅔ inches long.

COLOR: They range from nearly black to reddish brown.

Cactus Coreid

Chelinidea vittiger aequoris

Some species of prickly pear cactus (*Opuntia* spp.) are common plants on the deep sandy areas in longleaf pine forests. Though the many spines of the cactus don't make this plant very approachable, a closer inspection may reveal an insect that relies almost solely on the cactus for its existence.

SIZE: Described as short and stout, adults have a length of approximately ½–¾ inch.

COLOR: Wing margins are orange/yellow. Wings and head are black, with a distinctive orange/yellow stripe on the head. The pronotum (plate area on back behind head) is also orange/yellow.

BEHAVIOR: Some consider this a cactus pest. Both the adults and nymphs feed on the prickly pear pads and sometimes on the fruit. Adults hibernate on the underside of the pads starting in December and emerge around February. Eggs are laid on cactus spines. Light, circular spots on the cactus pads are signs of feeding.

ADDITIONAL NOTES: A similar species, the leaf-footed cactus bug (*Leptoglossus phyllopus*), can also be found feeding on prickly pear cactus (especially the fruit). It is believed to have been introduced to the southeastern United States on nursery stock.

Adult cactus coreid.

Large Milkweed Bug

Oncopeltus fasciatus

Many different species of milkweed plants (*Asclepias* spp.) are found throughout longleaf pine forests. Most people associate butterflies with milkweed. However, many insects, including milkweed bugs, rely on milkweeds for part or all of their life cycle.

SIZE: These are midsized insects, ½–¾ inch long.

COLOR: The body is alternating reddish orange and black. The head and rear are black and diamond shaped, while a black band runs across the middle of the back.

BEHAVIOR: These are often found in small groups feeding primarily on or near the seed pods of milkweed plants. Females lay eggs around the seed pods. With their piercing mouthparts, they consume the white sappy liquid of milkweeds. This liquid is toxic to many animals.

ADDITIONAL NOTES: The orange and black of the large milkweed bug is similar to the coloration of monarch butterflies. Like the large milkweed bug, these butterflies also ingest milkweed plants as caterpillars and store the toxic compounds. Many bird species have associated this orange/black with a bitter taste that results in them vomiting up their meal. Because they share the same colors, the milkweed assassin bug (*Zelus longipes*) is often mistaken for the milkweed bug.

Large milkweed bug.

Antlion

Myrmeleon carolinus

Although most people associate antlions with the shallow cone-shaped pitfall traps, the term "antlion" can actually describe many different species of like-kind insects in longleaf pine forests. Some build these pitfall traps and some spend their entire life up in a tree. *Myrmeleon carolinus* is one of the most common. Though the size of the pitfall trap and slight distinguishing features of the larvae and adults are different among species, usually only experts can tell them apart. The larvae are the most easily recognized form of this insect; they make the pitfall traps. They have a stumpy body, flat head, and large sickle-shaped pinchers in the front. Adults are active at night and are rarely seen, as they are well camouflaged in the day. Adults resemble dragonflies with four narrow, veiny transparent wings.

SIZE: Larvae are generally under ⅔ inch long. Adults can grow to about 1½ inches long.

COLOR: Head and jaws of the larvae are a distinguishing brown. The remainder of the body is dull tan/gray. Adults are also well camouflaged, with a dusky body and transparent wings.

Antlion larvae staged for a photograph but otherwise rarely seen out of cone-shaped pit (family Myrmeleontidae, unknown species).

Shallow cone-shaped pits are telltale signs of antlions.

BEHAVIOR: The larvae make their pitfall traps in open sandy areas. An ant may find its way into this trap and slide to the bottom and into the waiting jaws of the antlion. The jaws clamp down onto the ant and inject saliva, liquefying the ant's insides. When the meal is finished (about 2 hours after capture), the ant's carcass is flipped outside of the pit.

ADDITIONAL NOTES: The larvae can move only backward.

Palmetto Tortoise Beetle

Hemisphaerota cyanea

Also known as the Florida tortoise beetle, this small beetle is found in areas with abundant palms. They are often found on the fronds of saw palmetto (*Serenoa repens*) but can also be found on the other palm species common in the Southeast.

SIZE: These beetles are under ½ inch long.

COLOR: Adults are glossy black to iridescent blue, with yellowish tan antennae.

BEHAVIOR: These beetles have unique defensive tricks. Adult beetles use a combination of adhesive "foot" pads and secreted oil to clamp their shell down on the palm leaf, making it exceptionally difficult to remove them. The

Adult palmetto tortoise beetles.

Palmetto tortoise beetle frass on palmetto.

larvae of this beetle hide in a basket of their own feces (frass), protecting them from many predators.

ADDITIONAL NOTES: They are called tortoise beetles because of their shape and not because they have anything to do with gopher tortoises.

Scorpions

Southern Unstriped Scorpion (*Vaejovis carolinianus*)
Hentz Striped Scorpion (*Centruroides hentzi*)

Many people associate scorpions with deserts and dry environments, but some scorpions are also found in longleaf pine forests. The term "woods scorpion" is sometimes used by locals and likely describes one of two species: the southern unstriped scorpion or the Hentz striped scorpion. They are both relatively secretive arachnids that are often found under places like rocks or logs.

SIZE: Adults average 2¼–3 inches long.

COLOR: The largest visual difference between the two species is their color patterning. The Hentz striped scorpion is brown overall, with a yellowish stripe down the midline and sometimes down each side of the thorax. The southern unstriped scorpion lacks these yellow stripes and is overall reddish to brown.

Southern unstriped scorpion.

Hentz striped scorpion.

BEHAVIOR: Woods scorpions are not aggressive but will defend themselves. Some species' venom helps to digest their prey. Others have neurotoxins that immobilize their prey in order to disassemble them into more manageable portions without much of a fight. They typically eat crickets, spiders, roaches, and beetles. They have also been known to eat other scorpions.

ADDITIONAL NOTES: Slightly venomous to humans, the sting of a woods scorpion can cause local pain and swelling.

Grasses

Wiregrass

Aristida stricta and Aristida beyrichiana

Wiregrass is an important bunchgrass that provides a litter layer for burning, which is essential for a healthy habitat for plants and animals alike. Pine needles that become caught in the tangle of grass leaves provide fuel for fire to burn.

DESCRIPTION: This dense, perennial bunchgrass bears long and wiry leaves that grow low to the ground. The plant usually does not exceed 3½ feet in height. Another common name for this species is pineland three-awn, named for its seeds that bear three hairlike awns (a narrow bristlelike attachment at the end of the seed). Wiregrass produces abundant viable seed only when burned during the growing season months of April through July.

HABITAT: Wiregrass grows in a variety of frequently burned and open longleaf pine forests without a recent history of cultivation. It is found on sites ranging from very dry sandhills to very wet flatwoods and savannas. *Aristida stricta* occurs in North Carolina and the northern portion of the South Carolina coastal plain. The wiregrass growing in southern South Carolina, Georgia, Florida, Alabama, and Mississippi is known as *Aristida beyrichiana*. It is never found with longleaf pine in the mountains or from Central Mississippi to Texas.

ADDITIONAL NOTES: Wiregrass provides nesting sites for Northern Bobwhite and other ground-nesting birds. Gopher tortoises will forage on wiregrass, though it is of minor nutritional value for them. Seeds are

Longleaf stand with solid carpet of blooming wiregrass.

Wiregrass seed stalk.

occasionally eaten by songbirds. This grass supported much of early open-range grazing by Pineywoods cattle. However, with little nutritional value, it was speculated that roughly 2–4 acres of wiregrass range was required to support one cow annually. Other common species include arrowfeather three-awn (*Aristida purpurascens*) and woolysheath three-awn (*Aristida lanosa*).

Bluestems

The bluestems are important grasses in longleaf pine forests. They are generally tufted or bunched perennial species that occur across the range of longleaf pine and within a variety of habitats. These commonly occurring grasses provide fuel for fires and also benefits for wildlife, like food and cover (nesting and protection from predators).

Splitbeard Bluestem (*Andropogon ternarius*)

DESCRIPTION: This grass grows to a height of 2½–3¼ feet. The paired flowering clusters have silvery-white hairs and are attached to the main grass stem by stalks measuring up to 2 inches long.

Splitbeard bluestem.

Broomsedge.

HABITAT: Bluestem is found in dry, sandy soils throughout the longleaf range and is especially common in the montane longleaf pine habitat.

ADDITIONAL NOTES: Seeds are eaten by songbirds and Northern Bobwhite. White-tailed deer browse on this grass when more edible foods are less available.

Broomsedge *(Andropogon virginicus)*

DESCRIPTION: This common bunchgrass grows up to 4 feet tall. The leaves overlap and are flattened at the base of the plant. The blades of the leaf are usually hairy and measure 10–16 inches long and up to ½ inch wide. Broomsedge differs from the other bluestems in the number of racemes (flowering stalks) that grow in each inflorescence: each inflorescence can have 2–4 racemes instead of only 2.

HABITAT: Broomsedge is a typical ruderal or old-field species that likes disturbed sites and is usually one of the first to pioneer an area. It can be found growing in a variety of habitats though, including grasslands, pastures, open forests, and roadsides.

ADDITIONAL NOTES: Because of its growth habit, this grass provides good cover and nesting habitat for birds and small mammals. The seeds can provide a food source for Northern Bobwhite and songbirds. Similar species include Elliott's bluestem (*Andropogon gyrans*) and bushy bluestem (*Andropogon glomeratus*).

Big Bluestem *(Andropogon gerardii)*

DESCRIPTION: As its name implies, big bluestem is a large bunchgrass that can grow up to 8 feet tall. It is the largest of all the bluestem species for this region. The leaf blades can be as big as 2 feet long by ½ inch wide. Unlike many of the other bluestem species, the big bluestem has a brown inflorescence instead of silky and white. This plant grows from rhizomes, so you will normally see it growing in clumps.

HABITAT: This has a wide distribution in North America. It is one of the common grass species that occur in the central plains of the United States. In the Southeast, you will find this species growing in piney woods with more soil moisture—hillside bogs, pine savannas, and flatwoods.

Big bluestem flowering stem.

Big bluestem.

Little bluestem.

ADDITIONAL NOTES: Because of its growth habit, this grass provides good cover and nesting habitat for birds and small mammals. The seeds can provide a food source for Northern Bobwhite and songbirds.

Little Bluestem *(Schizachyrium scoparium)*

DESCRIPTION: This grass grows only up to 4 feet tall. The flowering cluster is arranged singly instead of doubly as in the other bluestem species. Leaves will turn purplish red in the fall.

HABITAT: Little bluestem grows well in a wide variety of habitats, including old fields, open forests, and prairies. This is the dominant grass in areas within the longleaf range where wiregrass does not occur.

ADDITIONAL NOTES: It provides good cover for birds and small mammals. Seeds are eaten by a variety of birds, especially the native sparrows. Northern Bobwhite nest adjacent to these clumps, with overhead cover provided by the grass. Pine Warblers use dead material in nest construction in adjacent trees. Another common bluestem species is slender bluestem (*Schizachyrium tenerum*).

Toothache Grass
Ctenium aromaticum

This perennial grass is easily recognizable when in flower or fruit but can be confused with Florida dropseed if not in flower. Historically, the roots and lower culms were used as a treatment for toothaches due to the presence of an analgesic (pain-relieving chemical), a property that gives this grass its common name. The weak effect, and difficulty of identifying most grass species, makes using this plant for any medicinal purpose risky and is not recommended by the authors.

DESCRIPTION: This perennial bunchgrass forms dense clumps with a height of 2–3½ feet. The leaves are dark green on top and lighter green on the bottom. The flowering cluster looks like a comb and curls as seeds mature.

Curled seed head of toothache grass.

Toothache grass stand in bloom.

HABITAT: This is a common grass in wet to moist pine flatwoods, savannas, prairies, and pitcher plant bogs.

ADDITIONAL NOTES: The bunched structure of this grass provides good cover for wildlife, while seeds can be foraged by a variety of grassland birds such as Bachman's Sparrows.

Purple Lovegrass

Eragrostis spectabilis

DESCRIPTION: This attractive fall-blooming tufted grass is fairly common throughout longleaf pine forests. It is a relatively short, bunched perennial with hairy leaf blades. The open-panicle inflorescence has purple-tinted spikelets, giving the flowering stem a purple cloud-like appearance.

HABITAT: This grass is often seen growing on roadsides and is especially noticeable when the plant is in flower. It also grows in sandy fields, pastures, and open woods throughout the range of longleaf pine.

ADDITIONAL NOTES: Birds use the seeds of purple lovegrass as a food source.

Purple lovegrass.

Hairawn Muhly Grass

Muhlenbergia capillaris

DESCRIPTION: This robust, perennial bunchgrass can reach 3½ feet tall when in flower. This species is usually found alongside wiregrass and can be confused with that species because of its wiry leaves. However, muhly grass has less pronounced tufts of hairs that are conspicuous at the base of wiregrass leaf blades. In addition, hairawn muhly lacks pronounced awns on lemma (sheaths covering seed-producing structures), while wiregrass awns

Hairawn muhly grass.

are large and pronounced. The flowering structure of muhly grass is also very different from wiregrass. Muhly blooms in the fall and has a large, branched flowering stem that is pinkish purple. This species is widely used in the horticulture industry because of its beautiful pink flowering stems and its ability to grow in tough growing conditions.

HABITAT: Muhly grass grows in a wide range of sites, including seeps, marshes, stabilized sand dunes, pine savannas, and flatwoods. It occurs throughout the range of longleaf pine.

ADDITIONAL NOTES: This large bunchgrass provides good cover for wildlife, and a variety of bird species eat the seeds. Savanna hairgrass (*Muhlenbergia expansa*) is a closely related species with similar characteristics; it is closely associated with longleaf pine ecosystems in the coastal plain.

Blackseed Needlegrass

Piptochaetium avenaceum

DESCRIPTION: This densely tufted perennial grass has relatively short, wiry leaves that measure up to 20 inches long. The leaves are dark green and have a wiry appearance because the leaf blades are tightly curled inward. This is one of the relatively few grass species in longleaf forests that bloom in spring. The flowering stem of blackseed needlegrass is arranged in loose panicles with

Blackseed needlegrass.

multiple spikelets (grass flowers). The spikelets have long awns that will twist and contort as they mature.

HABITAT: This grass species occurs in dry, open hardwood forests, longleaf pine forests, dry to mesic mixed pine/hardwood forests, and sandy woods. It is a dominant grass in montane and piedmont longleaf forests and has a wide distribution, growing beyond the boundaries of the longleaf range.

ADDITIONAL NOTES: This species has limited wildlife food value; however, the seeds are sometimes consumed by songbirds. Its seeds attach to fur and clothing and are dispersed by unwitting mammals. The structures that allow this form of dispersal is how the plant gets its common name of needlegrass. It is an important fuel source for prescribed fires in parts of the longleaf range, where it is a dominant groundcover component. These prescribed fires are essential for maintaining quality wildlife habitat.

Lopsided Indiangrass

Sorghastrum secundum

This perennial bunchgrass is increasingly being used in native understory restoration of longleaf pine habitat across the central portion of the region. This species provides wildlife benefits, provides fuel for prescribed fires, and is aesthetically pleasing.

Lopsided Indiangrass.

DESCRIPTION: The height ranges from 3 to 6 feet. One major identifying characteristic for the Indiangrasses is the presence of a long membranous ligule at the leaf blade base measuring up to ½ inch long. The flowering cluster is arranged on one side of the stalk (lopsided) and the individual spikelets have long, twisted bristlelike ends.

HABITAT: It grows in well-drained soils of upland sites in Florida, Georgia, and Alabama.

ADDITIONAL NOTES: This bunchgrass provides good dry forage during the winter. It also provides cover and food sources for many wildlife species. Seeds are eaten by songbirds and small mammals. It is also a larval host plant for several skipper moth species. Two other species of Indiangrass can be found in longleaf forests: yellow Indiangrass (*Sorghastrum nutans*) and slender Indiangrass (*Sorghastrum elliottii*)

Pineywoods Dropseed

Sporobolus junceus

This commonly occurring perennial bunchgrass is usually found growing with wiregrass. Pineywoods dropseed and wiregrass are two species that can sometimes be confused when not in flower.

DESCRIPTION: This is a relatively small bunchgrass that reaches a maximum height of 3 feet when in flower. The leaves are wiry and blue/green. The grass blooms in the fall and the flowering clusters are pyramid shaped.

HABITAT: It is found in open, dry sites in rolling hills, longleaf pine habitat, drier flatwoods, and sandhills. This species is found throughout the range of longleaf pine.

ADDITIONAL NOTES: The structure of the plant provides good cover and the foliage is eaten by white-tailed deer. Another species of dropseed that can be found in wet savannas and flatwoods is Florida dropseed (*Sporobolus floridanus*).

Pineywoods dropseed.

Beaked panicgrass.

Beaked Panicgrass

Panicum anceps

This is one of the many *Panicum* species that live in the longleaf ecosystem. The panicgrasses and the closely related witch grasses (*Dicanthelium* spp.) are common in disturbed areas and are generally one of the first grasses you see colonizing these areas.

DESCRIPTION: This perennial grass grows from scaly stems called rhizomes that grow underground. These stems spread and create large clumps of shoots. When in flower, the plant can reach 3 feet tall. The flowering stem is arranged in a triangular panicle and measures 3–5 inches across.

HABITAT: It is found in open forests, forest openings, and along forest margins.

ADDITIONAL NOTES: This grass species produces large amounts of seed that are a food source for many bird species. Switchgrass (*Panicum virgatum*) is similar to beaked panicgrass but is generally a much larger plant and occurs in sites with higher soil moisture.

Forbs

Bracken Fern

Pteridium aquilinum

Pteridium is a wide-ranging genus of ferns that occurs around the globe. The species that occurs in the southeastern United States is very common and can be found growing in most longleaf habitats. It is especially common in longleaf stands with a history of fire in the dormant season. The dense growth of the fronds can form a groundcover canopy that will shade out other herbaceous species. In addition, bracken fern is known to reduce competition by releasing harmful chemicals to other plants, a strategy known as allelopathy.

DESCRIPTION: Bracken fern is a perennial fern with stiff, dark green fronds that arise from underground rhizomes. Each frond is widely triangular and can measure up to 2 feet long by 2 feet wide. They turn green early in the season and turn brown by early fall. As this is a fern, it produces spores on the underside of the frond leaflets.

Bracken fern.

HABITAT: This species is ubiquitous in the longleaf range and can be found in most habitats where longleaf occurs.

ADDITIONAL NOTES: Bracken fern can provide cover for wildlife but can also be toxic to livestock.

Pipeworts

Hatpins (*Eriocaulon decangulare*)
Common Bog Button (*Lachnocaulon anceps*)

The group of plants commonly called pipeworts are often found growing in wet areas within longleaf pine habitats like flatwoods and savannas. They are also known as hatpins, a name they get from the similarity of the flowering stem to the pins once used to secure hats to a person's hair. The bog button is the smaller cousin to the larger hatpin that is easily seen rising out of the vegetation in pitcher plant bogs. Both species respond quickly with new growth after a fire.

DESCRIPTION: These species in the pipewort family grow from rhizomes and tend to form dense groupings of plants. Hatpins are generally more robust and larger overall than the bog buttons, with stems reaching a maximum height of 32 inches compared to 10 inches for bog buttons. The leaves are only basal for both species. Hatpins have leaves that appear segmented with air spaces within and measure up to 12 inches in length and ½ inch in width. The linear leaves of the bog button are somewhat delicate, measuring up to 3 inches long by ¼ inch wide. The flowering stems grow out of the basal leaf rosettes. The flowering stalk of hatpins is smooth, whereas the bog button stems are covered in scattered long hairs.

HABITAT: Pipeworts grow in wet areas often in association with pitcher plants. They are found in these types of habitats throughout the longleaf range.

ADDITIONAL NOTES: Other common species that may be encountered in bogs and are similar in size to the common bog button include Small's bog button (*Lachnocaulon minus*) and yellow hatpin (*Syngonanthus flavidulus*).

Hatpins.

Common bog button.

Lilies

Native lily species found within longleaf ecosystems are often extremely showy. Most are not common, but their bright colors and large flowers make them hard to miss.

Pine Lily (*Lilium catesbaei*)

DESCRIPTION: This perennial grows from a bulb. The stems can be as tall as 2½ feet and have alternate leaves that are mostly present on the middle and upper part of the stem. The stem leaves are 1½ inches long by ½ inch wide and are narrowly elliptical. Each stem has only one flower that faces upward (as opposed to downward nodding like other lilies) with petals and sepals that are spreading and curved backward. Both the sepals and petals are reddish orange, with bases that are yellow with brownish purple dots.

HABITAT: This lily grows in wet savannas and flatwood areas that have recurrent fire. Often found in wet areas with bog buttons, pitcher plants, and wiregrass.

Panhandle Lily or Pot of Gold Lily (*Lilium iridollae*)

DESCRIPTION: Like the pine lily, the panhandle lily is also a perennial that grows from a bulb, but each plant will have several bulbs connected by underground stems called rhizomes. This species is much taller. The stems can grow to be over 6 feet tall. The leaves grow alternately on the lower portion of the stem and then are whorled on the upper portion. The leaves are larger and more rounded than the pine lily, measuring approximately 3 inches long and 1 inch wide. Each stem usually has only one flower but can sometimes have multiple. These flowers are nodding, and the yellowish orange petals with brown spots are curved backward. The long stamens hang down from the flower.

HABITAT: This has a narrow range, primarily in wet savanna habitat. It can only be found growing in swamps, bogs, and seepage slopes within longleaf forests in the Florida panhandle and in Alabama. They are frequently located in ecotone areas (area between two habitats) like wet savannas and unburned habitat (sometimes called baygalls).

Pine lily.

Panhandle lily.

ADDITIONAL NOTES: Several native lily species resemble the horticultural varieties many people grow in their gardens. Besides attracting people, the flowers are also beneficial for hummingbirds, butterflies, and bees.

Orchids

Many people are surprised to learn about the native orchids of the Southeast. The longleaf forest is home to quite a few of these colorful and exotic-looking plant species. They are commonly found in wet savanna habitats and are closely associated with pitcher plant bogs that are maintained with frequent prescribed fire. The species described here are some of the more common orchids found in healthy longleaf habitats.

Grass Pink (*Calopogon tuberosus*)

DESCRIPTION: This orchid is single stemmed and arises from a single grass-like leaf (thus the common name). Each flowering stem holds several vibrant magenta flowers that bloom in order from the bottom of the arrangement to the top of the stem. The bearded lip of the flower is located on the top of the flower. This species flowers in spring to early summer.

HABITAT: These are found throughout the entire range of longleaf pine. They grow in wet savannas, seepage slopes, and pitcher plant bogs.

Orange Fringed Orchid (*Platanthera ciliaris*)

DESCRIPTION: This orchid is single stemmed but has multiple narrow leaves, with the largest at the bottom of the stem (8 inches); leaves get progressively smaller up the stem. The flowers grow in a dense cluster at the top of the stem and are the color of orange sherbet. The lip of these beautiful flowers is heavily fringed. This orchid blooms late summer.

HABITAT: This species is found throughout the entire range of longleaf pine. They grow in wet savannas, seepage slopes, and pitcher plant bogs.

Grass pink.

Rose pogonia.

Orange fringed orchid.

Rose Pogonia (*Pogonia ophioglossoides*)

DESCRIPTION: This spreads vegetatively through root runners, resulting in clusters of single stems that hold a single flower. The stems have one ovate leaf at the mid-stem. The flower emerges from a bract that resembles a leaf at the top of the stem. These fragrant flowers are a pale pink with 3 narrow sepals, 2 petals, and 1 bearded lip that projects forward with white and yellow bristles. This species flowers in spring to early summer.

HABITAT: These are found throughout the entire range of longleaf pine. They grow in wet savannas, seepage slopes, and pitcher plant bogs.

ADDITIONAL NOTES: Orchids have unique flower structures that generally consist of four parts: sepals, petals, column, and a lip. Some species have very specialized relationships with insects and other animals that assist with pollination. Other species you might encounter growing along with these described species include crested fringed orchid (*Platanthera cristata*) and pale grass pink (*Calopogon pallidus*).

Milkweeds

These showy perennials can be found in a wide variety of habitats. These plants have unique flower structures and produce large amounts of nectar that attract all sorts of pollinating insects. Their most well-known visitor is the monarch butterfly, which lays its eggs on the plant. Some species are important larval food sources for butterflies. Besides the two species described here, other common species you may find in longleaf forests include clasping milkweed (*Asclepias amplexicaulis*), fewflower milkweed (*Asclepias lanceolata*), and Michaux's milkweed (*Asclepias michauxii*).

Sandhill Milkweed (*Asclepias humistrata*)

DESCRIPTION: This sprawling, multistemmed perennial has purple-tinted stems and thick leaves with pink veins. The distinctive milkweed blooms appear in the spring from April through May with flowers that range from pinkish white to pale purple. The seeds bear feathery attachments that help to disperse the seed when it ripens.

Sandhill milkweed.

HABITAT: This plant occurs on very dry, well-drained sandy soils. It can be found growing in sandhill habitats.

ADDITIONAL NOTES: These plants provide tremendous wildlife benefit to pollinators. Sandhill milkweed is a larval host plant for monarch and queen butterflies.

Butterfly Weed (*Asclepias tuberosa*)

DESCRIPTION: This upright, multistemmed perennial can reach a height of 2½ feet. Unlike many other milkweeds, butterfly weed does not have milky sap. It blooms from May through August, and the flower ranges from yellow to red. The seeds bear feathery attachments that help to disperse the seed when it becomes ripe.

Butterfly weed.

HABITAT: It is widely adaptable and occurs on dry to moist sites and open to semi-shaded habitats. It is a common plant throughout the range of longleaf pine.

ADDITIONAL NOTES: These plants provide tremendous wildlife benefit to pollinators, especially butterflies. This plant is a primary host plant for monarch butterflies.

Soft Greeneyes

Berlandiera pumila

This yellow-flowering member of the sunflower family blooms in spring and early summer. It is found growing throughout the longleaf range from South Carolina to Texas.

DESCRIPTION: It can grow to a height of 3 feet. It has oval, toothed leaves that are oppositely arranged on the stem. The flowers that have green centers generally grow singly at the end of long stalks.

HABITAT: This plant inhabits dry longleaf forests such as those in the rolling hills region.

ADDITIONAL NOTES: The flowers are a good pollen source for pollinating insects. Members of this genus are sometimes referred to as chocolate flowers due to their scent, which resembles the common confectionary ingredient.

Soft greeneyes.

Golden Aster

Chrysopsis mariana

These are among the many yellow-flowering plants that are so abundant in the fall landscape of the longleaf pine forests. These members of the sunflower family are found throughout the longleaf range and occur in a variety of habitats.

DESCRIPTION: This species can grow to a height of 2½ feet. The stems and leaves have silky white hairs. The bright yellow flowers that appear from summer into fall are approximately 1½ inches wide.

HABITAT: The seeds of golden aster are wind dispersed and establish well on disturbed sites. This plant can be found in new forests, forest openings, roadsides, and rights-of-way.

Whorled coreopsis.

Golden aster.

ADDITIONAL NOTES: The foliage of this species is eaten by white-tailed deer and rabbits. Two other similar species of golden asters that you may find growing with longleaf are cottony golden aster (*Chrysopsis gossypina*) and pineland golden aster (*Chrysopsis latisquamea*).

Whorled Coreopsis

Coreopsis major

This member of the sunflower family is often found flowering in the summer through early fall in longleaf pine forests. It has bright yellow flowers and is also known as greater tickseed.

DESCRIPTION: The stem of this perennial herbaceous plant can grow up to 3 feet high. The stalkless leaves are oppositely arranged on the stem and are so deeply dissected as to appear whorled on the stem. This plant flowers throughout the summer. Its typical sunflower-like flowers can be up to 3 inches wide with yellow ray and disc flowers.

HABITAT: This is one of many species of tickseed that occur in longleaf systems. This species is common in dry woodland areas in upper coastal plain and piedmont regions from Virginia south to Florida and west to Louisiana.

ADDITIONAL NOTES: *Coreopsis* species are attractive nectar sources for native pollinators. Several other species that you may see include lobed tickseed (*Coreopsis auriculata*), star tickseed (*Coreopsis pubescens*), and coastalplain tickseed (*Coreopsis gladiata*).

Roundleaf Thoroughwort

Eupatorium rotundifolium

In the fall, members of the genus *Eupatorium* may look like a sheet of white lace across the landscape. Many different species of this plant type can be found in longleaf pine forests. Roundleaf thoroughwort is a distinctive species with distinct rounded leaves that aid in identification.

DESCRIPTION: This perennial plant species can grow to 3 feet tall. The plant is covered with soft hairs, the stem is branched, and the rounded, toothed leaves are arranged oppositely along the stem. This fall-blooming species has flat flowering clusters of tiny, white flowers.

HABITAT: This plant occurs in a wide range of soil moisture conditions. It can be found in pinelands, oak-hickory woods, savannas, and stream margins.

Roundleaf thoroughwort.

ADDITIONAL NOTES: Songbirds eat the seeds, white-tailed deer browse the foliage, and pollinators are attracted to the flowers. Other common species include yankeeweed (*Eupatorium compositifolium*), justiceweed (*Eupatorium leucolepis*), and smallflower thoroughwort (*Eupatorium semiserratum*).

Plants employ various dispersal mechanisms to maximize seedling recruitment, including (clockwise from top left) gravity, wind, ballistic, internal animal transport, and external animal transport.

Asters

This group of plants in the sunflower family is represented by a diverse assemblage of species. The majority of them bloom during the fall and it can sometimes be difficult to differentiate between species. Originally placed within the genus *Aster*, the species that are found in eastern North America were divided among six different genera based on their morphological characteristics: *Eurybia*, *Doellingeria*, *Ionactis*, *Oclemena*, *Sericocarpus*, and *Symphyotrichum*. The species described here are just a few of the many varieties that can be found growing in the longleaf pine forest.

Stiff-Leaved Aster (*Ionactis linariifolius*)

DESCRIPTION: Stiff-leaved aster is a multistemmed, herbaceous perennial with dark green leaves that are stiff and sharply pointed. The stems can reach a maximum height of 19 inches; the narrow leaves generally measure 1 inch long. In the fall, the flowers form on branches at the top of the stems. These flowers are composites and are comprised of bluish/purple ray flowers and yellow disc flowers.

HABITAT: This species can be found in a wide range of longleaf pine habitats but is most common in those sites with drier soils.

Silvery Aster (*Symphyotrichum concolor*)

DESCRIPTION: This perennial aster species is usually single stemmed, stands 2 feet tall, and has soft, light green leaves covered in silvery hairs. The leaves are sessile, measure approximately 1½ inches in length, and are closely arranged on the stem. Silvery aster blooms also in the fall, and the light purple flowers are arranged in a narrow cluster at the top of the stem.

HABITAT: This species can be found in a wide range of longleaf pine habitats.

Dixie White-Topped Aster (*Sericocarpus tortifolius*)

DESCRIPTION: This perennial aster species is one of the many white-flowering asters that bloom in the fall. It generally is composed of single stems that reach up to 3 feet tall. The stems and leaves are covered in short hairs. The rounded leaves also have sticky glands on their surface that make

Stiff-Leaved Aster.

Silvery Aster.

Dixie White-Topped Aster.

the leaves glisten slightly in the sun. The white flowers emerge in early fall. The ray flowers are usually fewer in number than the disc flowers.

HABITAT: This species can be found in a wide range of longleaf pine habitats but is most common in those sites with drier soils.

ADDITIONAL NOTES: Asters attract a wide variety of pollinators during their flowering season. Other common species not described here include rice button aster (*Symphyotrichum dumosum*), scaleleaf aster (*Symphyotrichum adnatus*), and Walter's aster (*Symphyotrichum walteri*).

Blazing Stars

The blazing stars are a very showy group of flowering plants that are fairly common throughout the longleaf pine range. Their flowers are highly attractive to pollinators and provide a diverse nectar source every fall.

Pinkscale blazing star.

Pinkscale Blazing Star (*Liatris elegans*)

DESCRIPTION: Like most of the *Liatris* species, pinkscale blazing star grows from a hard, round rootstock. The stem is generally unbranched, hairy, and has long leaves (up to 12 inches long) at the base of the stem, with much reduced leaves in the upper portion. This species can be easily distinguished from the other species of blazing star by the color of the flowers and the showy petallike bracts. Flowers are arranged in clustered heads at the top of the stem and are white to pale pink. Each flower cluster is surrounded by pink bracts (small, modified leaf under the flower). Blooming occurs in the fall.

HABITAT: This grows in a variety of growing conditions and can be found in dry to mesic longleaf pine habitats.

Shortleaf blazing star.

Shortleaf Blazing Star (*Liatris tenuifolia*)

This drought-tolerant species can survive in the driest habitats. It is among the more common blazing stars encountered in longleaf pine ecosystems.

DESCRIPTION: This perennial species grows from a hard, round rootstock. Until the flowering stalk bolts in late summer, the plant persists as a cluster of low-growing grasslike leaves. It reaches a maximum height of 6 feet. The purple flowerheads grow on a long, dense, spikelike flowering stalk.

HABITAT: This grows in dry turkey oak sandhills and scrub.

ADDITIONAL NOTES: The blazing stars attract many types of native pollinators, especially butterflies. The seeds also provide a good food source for songbirds. Other common species of *Liatris* that can be found in longleaf habitats include shaggy blazing star (*Liatris pilosa*), prairie blazing star (*Liatris pycnostachya*), and scaly blazing star (*Liatris squarrosa*).

Silkgrass flower and stalk.

Silkgrass

Pityopsis graminifolia

This member of the sunflower family can be found throughout the range of longleaf pine. It is very common and found in a variety of habitats and soil conditions. The silvery color of the leaves makes this plant very distinctive.

DESCRIPTION: This herbaceous perennial resembles grass when not in flower. The plants form clumps of narrow silvery leaves. In late summer, silkgrass forms showy masses of small yellow daisy-like flowers.

HABITAT: It can be found in areas where young trees are being cultivated, open forests, forest openings, and right-of-way areas.

ADDITIONAL NOTES: This widespread species is an important food for gopher tortoises. There are several similar species that occur across the longleaf range. These include Carolina silkgrass (*Pityopsis aspera*), grass leaf goldaster (*Pityopsis oligantha*), and Taylor County silkgrass (*Pityopsis pinifolia*).

Black-eyed Susan.

Black-Eyed Susan

Rudbeckia hirta

This is easily one of the most recognizable wildflowers here in the southeast. Black-eyed Susan is widely used in backyard gardening and wildflower seed mixes.

DESCRIPTION: This short-lived perennial has stems and leaves covered with rough hairs. The large, bright yellow flower heads that are formed throughout the summer produce large amounts of seed.

HABITAT: It can be found in areas of planted pines, open forests, forest openings, and rights-of-way.

ADDITIONAL NOTES: The abundant seed produced by this colorful species provides a good food source for songbirds. Other common species include orange coneflower (*Rudbeckia fulgida*) and Mohr's coneflower (*Rudbeckia mollis*).

Anisescented Goldenrod

Solidago odora

The licorice or anise odor the leaves give off when crushed gives this plant its name. It has often been used for brewing an aromatic tea, and there is some evidence that the plant was used by Native Americans for medicinal purposes.

DESCRIPTION: It reaches a height of approximately 2 feet. The leaves are narrow. The yellow flowerheads emerge in the fall on a panicle.

HABITAT: This species is found throughout the eastern portion of the United States and grows in a variety of woodland habitats, ranging from longleaf pine sandhills to mixed pine/hardwoods.

ADDITIONAL NOTES: Seeds of this plant are consumed by small mammals and songbirds, the leaves are eaten by various wildlife species, and the flowers are attractive to pollinators. The nectar of goldenrod attracts the nonnative lovebug (*Plecia nearctica*). Other common species that occur in longleaf habitat include Canada goldenrod (*Solidago canadensis*), wrinkleleaf goldenrod (*Solidago rugosa*), and wand goldenrod (*Solidago stricta*).

Anisescented goldenrod.

Tall Ironweed

Vernonia angustifolia

DESCRIPTION: This perennial can grow to a height of approximately 3 feet. The plant has no basal leaves growing at the lowest part of the stem but has many linear-shaped leaves that are arranged alternately along the stem. These leaves measure 2–5 inches long and ¼ inch wide; the edges tend to roll under. The purple disk flowers appear during the summer from June to September. The flower clusters are arranged in a terminal flat-topped panicle.

HABITAT: It occurs in the coastal plain of the southeastern United States from North Carolina south to Florida and west to Mississippi. Naturally occurring tall ironweed plants are typically found growing in the dry soils of longleaf pine rolling hills.

ADDITIONAL NOTES: This is an attractive nectar plant for native pollinators.

Tall ironweed.

Eastern Prickly Pear

Opuntia mesacantha

This native cactus species can be found in some of the region's driest longleaf habitats. Pads are heavily armed with spines and glochids (small hairlike spines with barbs), making them unpleasant to step on or handle.

DESCRIPTION: This perennial species mainly grows prostrate (laying down) along the ground. It is much smaller in stature than the erect prickly pear (*Opuntia stricta*), which can sometimes be found around old homesites and in coastal dunes. The stems are segmented into flattened pads that can be up to 7 inches long and 3 inches wide. Each pad has a long spine and scattered tufted glochids that break off when touched. This plant blooms in early summer with bright yellow flowers and produces fleshy fruits with numerous seeds.

HABITAT: It occurs throughout the range of longleaf and thrives in sites with dry, sandy soil.

ADDITIONAL NOTES: The succulent pads and fleshy fruits are a favorite food of gopher tortoises and prickly pear cactus bugs. White masses are common on the surface of the cactus. These masses are produced by the cochineal scale (*Dactylopious* spp.). When these scale (bugs) are crushed, they turn a brilliant red.

Eastern prickly pear.

Sandhill Dawnflower

Stylisma patens

This is a member of the morning glory family. It is related to the familiar cultivated and weedy morning glories but is much smaller. This species is common in the longleaf pine forest and can be found from North Carolina to Mississippi.

DESCRIPTION: This perennial is a nonclimbing, trailing vine with stems that can be as long as 4 feet. The soft leaves are elliptical and grow to 2 inches long. The white, funnel-shaped flowers are formed throughout the summer.

HABITAT: It inhabits dry to very dry longleaf pinelands and sandhills, especially where fire is used regularly.

ADDITIONAL NOTES: The flowers provide benefits for pollinators. Northern Bobwhite and songbirds may consume the seeds.

Sandhill dawnflower.

Tread Softly

Cnidoscolus stimulosus

Tread softly is aptly named because of the stinging hairs covering the entire plant. You learn quickly not to touch this prickly member of the spurge family. Each stinging hair is filled with a chemical that causes a burning sensation and rash for many people.

DESCRIPTION: This species grows up to 1½ feet and has palm-shaped leaves on short stalks. The showy white flowering parts are clustered terminally on the plant.

HABITAT: It occurs in dry, sandy soil of longleaf pine habitat in the sandhills and rolling hills regions. It is also found on rights-of-way, old fields, and sand dunes.

ADDITIONAL NOTES: The seeds are infrequently eaten by Northern Bobwhite and songbirds. White-tailed deer will eat the young shoots before the stinging hairs emerge. However, it is an old wives' tale that planting tread softly around your garden will repel deer. Texas bull nettle (*Cnidoscolus texanus*) is a similar but much larger species that occurs in the western portion of the longleaf range.

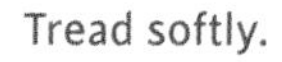
Tread softly.

Healing Croton

Croton argyranthemus

Crotons are members of the spurge family. Like many other plants in this family, healing croton has milky sap in the stems and leaves that is meant to deter animals from eating the plant.

DESCRIPTION: This short-lived perennial has distinctive copper-colored scales covering the main stem and the underside of the leaves. The plant is usually branched and can reach a maximum height of 2 feet. It blooms during spring and early summer with silvery-white flowers. The fruit is a capsule that splits open at maturity to release the seeds.

HABITAT: This species has a wide distribution and occurs throughout the longleaf range. It can be found growing in dry habitats with deep sandy soils.

ADDITIONAL NOTES: It is also an important plant for pollinators. Bees are the primary pollinator, and the goatweed leafwing butterfly (*Anaea andria*) uses it as a larval food plant.

Healing croton.

Wooly Croton

Croton capitatus

DESCRIPTION: This annual can grow to a height of over 3 feet. The plants are covered in velvety hairs, with the stems and underside of the leaves orange-brown and the upper side of the leaves white. The leaves are alternately arranged on the stem and are heart shaped at the base. The separate male and female flowers appear mid- to late summer on the plant and share the characteristics of other spurge family plants. The flowers lack petals but are surrounded by hairy bracts that give the flowers a wooly appearance. The fruit is a three-seeded capsule that splits open at maturity to release the seeds.

HABITAT: This species has a wide distribution and occurs throughout the longleaf range. It can be found growing in disturbed sites, locations of pine cultivation, and pastures.

ADDITIONAL NOTES: The seeds of wooly croton provide a tremendous food source for doves and Northern Bobwhite. It is also an important

Wooly croton.

plant for pollinators. Butterflies utilize the plant as a nectar source and goatweed leafwing and gray hairstreak butterflies use it as a larval host.

ADDITIONAL NOTES: Other *Croton* species that may be encountered in longleaf habitats include doveweed (*Croton glandulosus*) and sand rushfoil (*Croton michauxii*).

Gopherweed.

Gopherweed

Baptisia lanceolata

This widespread member of the legume family occurs in sandy habitats here in the Southeast. It is also called false indigo. Like many plants in this genus, gopherweed is a long-lived perennial species that emerges early in the spring and produces flowers around April. The flowers are pollinated by bumblebees.

DESCRIPTION: This bushy perennial herb grows to a height of 1–3 feet. Like many other legume species, the leaf is made up of a cluster of three leaflets. This compound or trifoliate (groups of three) leaf has leaflets that are each 1–4 inches long. Stems emerge from the ground early in the spring and bear bright yellow flowers between April and May. The fruit is a black legume measuring ½–1 inch long that remains on the plant from June through November.

HABITAT: Gopherweed is a common plant within its range. Found in sandhills and open woods from South Carolina to Alabama and south to Florida, it generally prefers sites with sandy, well-drained soils.

ADDITIONAL NOTES: This species is a larval food plant for the wild indigo duskywing (*Erynnis baptisiae*). It is largely replaced by the similar-looking Carolina wild indigo (*Baptisia cinerea*) in North and South Carolina.

Nuttall's Wild Indigo
Baptisia nuttalliana

DESCRIPTION: This perennial grows to approximately 3 feet tall and like many other legume species has trifoliate compound leaves. The yellow blooms emerge in the spring from the leaf axils and mature into a rounded seed pod that is covered in downy hair. The seed pods will turn blue-black when they are fully mature in late summer.

HABITAT: This species grows in pinewoods and prairies in south-central Louisiana and southeastern Texas.

ADDITIONAL NOTES: It can be toxic to wildlife, though the flowers are beneficial for native bee species. Other common species include pineland wild indigo (*Baptisia lecontei*) and white wild indigo (*Baptisia alba*).

Nuttall's wild indigo.

Spurred Butterfly Pea
Centrosema virginianum

This vining legume can be found throughout the Southeast from Texas to Florida and north to New Jersey.

DESCRIPTION: It is a twining or trailing vine with trifoliate compound leaves. This species flowers during the summer. The large 1-inch purple flowers with spreading petals form narrow 2½- to 5-inch legume pods that contain up to 20 seeds. The seeds disperse by a method called ballistic dispersal: the dry seed pod explodes under pressure, throwing seeds many feet away from the plant.

HABITAT: This plant can be found in rolling hills, montane areas, and flatwoods. It is less common on drier sites such as sandhills.

ADDITIONAL NOTES: The seeds are eaten by Northern Bobwhite and songbirds. The foliage is eaten by white-tailed deer and gopher tortoises. Spurred butterfly pea is very similar to Atlantic pigeonwings (*Clitoria mariana*). The two can be distinguished in flower by a spur under the base of the petals in the spurred butterfly pea and the open petals of Atlantic pigeonwings that form "wings." The seeds of Atlantic pigeonwings are sticky and use that feature to aid in dispersal. Vegetatively, spurred butterfly pea is vine-like, as opposed to the non-vine-like growth habit of Atlantic pigeonwings.

Spurred butterfly pea.

Partridge Pea

Chamaecrista fasciculata

This common annual legume species can be found throughout the longleaf pine range. This plant, like most legumes, functions as a nitrogen fixer. Nitrogen fixation occurs by means of a symbiotic relationship with certain bacteria that live in the plant's root tissues, using structures called nodules. Due to this relationship, these plants and bacteria can take nitrogen from the air and return it to the soil. This species is widely used for wildlife plantings across the southeast.

DESCRIPTION: This upright species can reach 3–4 feet tall. The leaves are compound, and the arrangement of the leaflets resembles a feather. This plant blooms throughout the summer and produces abundant amounts of small, hard seed. The large yellow flowers with red centers measure 1–2 inches wide.

HABITAT: It is commonly found in disturbed sites, such as wildlife food plots, roadsides, and areas cleared by logging. It also grows well in areas of pine cultivation, old fields, and open forests.

ADDITIONAL NOTES: The seeds are eaten by Northern Bobwhite, Eastern Wild Turkey, and small mammals. Partridge pea is also fed upon by white-tailed deer and gopher tortoises. In addition, this is a host plant for butterfly larvae. Other similar species include the small flowered partridge pea (*Chamaecrista nictitans*) and the perennial Florida Keys sensitive pea (*Chamaecrista deeringiana*).

Partridge pea.

Rabbitbells

Crotalaria rotundifolia

This common legume species occurs throughout most of the longleaf range. It gets its name from its dry seed pods that make a rattling sound when shaken. Some may mistake this for a rattlesnake when they brush against fruiting plants.

DESCRIPTION: This low, spreading plant seldom exceeds 16 inches in height. This is one of the legume species with simple leaves. The leaves are fleshy and rounded. The yellow flowers emerge during the summer and produce an inflated pod that rattles when seeds ripen and separate from the inside of the pod.

Rabbitbells flower.

Rabbitbells seed pods.

Littleleaf ticktrefoil.

HABITAT: This can be found growing in dry longleaf pinelands and disturbed areas.

ADDITIONAL NOTES: The seeds are occasionally eaten by Northern Bobwhite, and gopher tortoises may eat the leaves. Rabbitbells can be confused with Pursh's rattlebox (*Crotalaria purshii*) but can be distinguished by the narrower shape of the leaf and the pronounced arrow-like wings on the stems of *Crotalaria purshii*. The nonnative, invasive showy rattlebox (*Crotalaria spectabilis*) can be especially problematic in old field habitats.

Littleleaf Ticktrefoil
Desmodium ciliare

There are many species of ticktrefoils that occur within longleaf habitats. Many are familiar with these plants because of the seed pods that attach to their clothing while walking in the woods late summer through fall. This group of plants can be difficult to identify because of their similar morphology, and several of the species can also hybridize.

DESCRIPTION: This perennial legume has an erect growth habit, with stems that can grow to 3 feet tall. Like many of the *Desmodium* species, the whole plant is covered in short, hook-shaped hairs. These hairs act as a natural Velcro that will stick to animals brushing by the plant. The leaves are compound, with three ovate leaflets. The typical pealike flowers are purple and can be seen from summer into fall. The seed pods (loments) ripen in the fall.

HABITAT: This species can be found growing in a wide range of habitats but is frequently found in rolling hills habitat with dry to more moist soil. They can also be found in fields and roadsides.

Sand Ticktrefoil
Desmodium lineatum

DESCRIPTION: This perennial legume has trailing stems that grow along the ground. The whole plant is covered in short, hooked hairs. The leaves are composed of three rounded leaflets. The flowering stems grow from the leaf axils, and the flowers are purplish. They bloom throughout the summer, with the seeds ripening in the fall. The seeds are formed within a loment made up of 2–3 segments. Each segment contains one seed.

HABITAT: This species can be found growing throughout the range of longleaf pine in dry, sandy upland habitats.

ADDITIONAL NOTES: Ticktrefoil seeds are an important food source for Northern Bobwhite, doves, Eastern Wild Turkey, other birds, and small mammals. In addition, the plant is eaten by white-tailed deer. Other common species include Florida ticktrefoil (*Desmodium floridanum*), smooth ticktrefoil (*Desmodium laevigatum*), and pineland ticktrefoil (*Desmodium strictum*).

Sand ticktrefoil leaves.

Sand ticktrefoil flowers.

Coral Bean

Erythrina herbacea

This is a distinctive-looking member of the legume family: its tubular red flowers and shrubby growth stand out against the more common yellow and purple flowers of this plant group. It is also very toxic to humans but is used on occasion as an ornamental plant around housing.

DESCRIPTION: This perennial has multiple stems that tend to sprawl from the base and grow up to 5 feet in length. The stems and leaves both have spines, and the leaves are compound, with 3 spade-shaped leaflets. These leaflets are quite large, measuring 4 inches long by 3 inches wide. Coral bean blooms with red tubular flowers arranged in spikes during late spring. The seed pod can grow to 4 inches long and when mature will split open to reveal the bright red seeds.

HABITAT: This coastal plain species occurs from North Carolina south to Florida and west to Texas. It is usually found growing in dry sandhill sites.

ADDITIONAL NOTES: Due to the color and shape of the flowers, coral bean is attractive to hummingbirds as a nectar source.

Coral bean.

Sandhill Milkpea

Galactia regularis

Several similar-looking milkpea species occur within the longleaf range. Sandhill milkpea is one of the easier ones to identify because of its growth habit, which is more sprawling and prostrate compared to its vining cousins. This species provides significant wildlife benefits as a food and nectar source.

DESCRIPTION: This trailing or twining vine has stems up to 10 feet long. The stem is covered with pressed hairs. The attractive pink flowers grow throughout the summer and are loosely clustered where the leaves grow from the stem.

HABITAT: This plant is found in a variety of longleaf pine and other mixed pine-hardwood forests in the Southeast.

ADDITIONAL NOTES: The seeds are eaten by Northern Bobwhite, songbirds, Eastern Wild Turkey, and small mammals. The foliage is eaten by gopher tortoises and white-tailed deer. Other common *Galactia* species include erect milkpea (*Galactia erecta*), soft milkpea (*Galactia mollis*), and eastern milkpea (*Galactia volubilis*).

Sandhill milkpea.

Bush Clovers

Lespedezas are an important group of plants that provide multiple benefits in the longleaf forest. These members of the legume family contribute greatly to the high diversity of groundcover plant species. The foliage is eaten by white-tailed deer and gopher tortoises. They are also one of the most important wildlife food plants for Northern Bobwhite, and seeds are eaten by numerous songbird species. Traditional quail management programs have historically used nonnative lespedeza species, but the native species of lespedeza are generally more nutritious and less weedy.

Hairy Lespedeza (*Lespedeza hirta*)

DESCRIPTION: This common herbaceous perennial can grow up to 6 feet tall. As the name implies, it is covered in spreading tawny hairs. The leaves are trifoliate, and the leaflets are oval. The white/pale yellow flowers grow in tight,

Silvery bush clover.

Hairy lespedeza.

rounded clusters at the base of the leaves from summer to fall. The fruits are a single, flattened legume.

HABITAT: This species inhabits a variety of dry pineland habitats throughout longleaf pine's range.

Silvery Bush Clover (*Lespedeza capitata*)

DESCRIPTION: This is similar in appearance to hairy lespedeza. They are roughly the same maximum height and width but differ in the shape of the leaflets and the color of the hairs covering the plant. Silvery bush clover has dense, silvery hair as opposed to tawny hairs. The leaflets are narrower, with an elliptical shape instead of ovate. The flowers are yellowish white and arranged in dense clusters on the upper portion of the plant. The fruit is a single-seeded pod.

HABITAT: This species occurs in moist longleaf pinelands throughout longleaf pine's range.

Planting for Wildlife

Landowners have been planting lespedeza to provide food for wildlife for many decades. However, instead of one of our many native species, Asian species such as bicolor lespedeza (*Lespedeza bicolor*), Chinese bush clover (*Lespedeza cuneata*), and Thunberg lespedeza (*Lespedeza thunbergii*) have been widely used in wildlife food plots, forage, or hay. These shrubby species produce abundant seed and cover for Northern Bobwhite (*Colinus virginianus*) but are reported to be less palatable and less nutritious than our native bush clovers. These Asian species can become a problem in portions of the longleaf pine range with clay soils, where they are invading natural woodlands and displacing many native plant species. The use of native lespedezas is currently being recommended as a better alternative for food plots, but availability of seeds and long establishment times present problems for large-scale adoption.

Bicolor lespedeza.

Lupines

Lupines are early spring-flowering plants that provide a beautiful purple to the spring landscape in mostly dry longleaf habitats. They are important in providing an early nectar source for many native pollinators.

Sundial Lupine (*Lupinus perennis*)

DESCRIPTION: This perennial spreads from creeping underground stems called rhizomes. It is often found growing in small patches. When in flower, sundial lupine reaches 2½ feet tall. The palmately compound (radiate out from a single point) leaves of sundial lupines are primarily basally arranged, smooth, and each leaf is made up of 7–11 short leaflets that measure ¼–1 inch long. The elongated cluster of flowers that emerge in the spring are a deep blue-purple color. The hairy legume fruits are formed over the summer.

Sundial lupine.

Sandhill lupine.

HABITAT: This grows in dry habitats such as sandhills and turkey oak scrub. Its distribution is fairly widespread across the eastern portion of North America from Canada, south to Florida, and west to Texas.

Sandhill Lupine (*Lupinus diffusus*)

DESCRIPTION: This biennial grows from a tap root and is often found growing in small patches. When in flower, the sandhill lupine measures up to 1 foot tall. The leaves of this species are very different from the sundial lupine. Sandhill lupine has a simple elliptic leaf measuring up to 4½ inches long that is covered in silvery hairs, giving it a blue-green color. In the spring, the sandhill lupine has blue-purple flowers that bear a creamy white spot on the standard (upper winged petal of a legume flower). The hairy legume fruits are formed over the summer.

HABITAT: Both sundial and sandhill lupine grow in dry habitats such as sandhills and turkey oak scrub. Sandhill lupine distribution is from North Carolina south to Florida and west to Mississippi.

ADDITIONAL NOTES: Lupines, in general, are widely used by native pollinators for nectar. Additionally, the sundial lupine plays an important role as a larval host plant for two butterfly species: frosted elfin butterfly (*Callophrys irus*) and the Karner blue butterfly (*Plebejus melissa samuelis*); the latter does not occur in the longleaf pine range.

Eastern Sensitive Briar
Mimosa microphylla

Notably, when the leaves of this common legume species are touched, the leaflets will close in response (thigmonasty). The pinkish purple flowers look like those of the mimosa tree (*Albizia julibrissin*).

DESCRIPTION: This plant has prickly stems that creep flat along the ground. As the growing season progresses, it will climb nearby plants to escape shade. The plant bears pink pom-pom flowers throughout the summer and produces thorny pods containing several seeds each.

HABITAT: This species can be found in dry longleaf pinelands, openings, or burned areas in mixed pine-hardwoods and old fields.

Eastern sensitive briar.

Dollarleaf.

ADDITIONAL NOTES: The seeds are eaten by Northern Bobwhite, songbirds, and gopher tortoises. The foliage is eaten by gopher tortoises and Eastern Wild Turkey.

Dollarleaf

Rhynchosia reniformis

This legume is common across much of longleaf pine's range and ecosystems. Like most legumes, it is a nitrogen fixer through a symbiotic relationship with bacteria and provides a valuable seed source for numerous birds and small mammals.

DESCRIPTION: This perennial plant species is different from many of its relatives by having simple leaves. It has relatively short stems 3–7 inches tall that are densely hairy. As the scientific name indicates, the leaf is kidney shaped; it has a wrinkled surface and can measure up to 2 inches wide. The bright yellow pea flowers of this species can be seen from June to September.

The flowers and subsequent pea pods are found on short stalks and are tightly clustered around the leaf axils.

HABITAT: This species occurs in sandhills, rolling hills, and flatwoods. It can be found in most of the longleaf range from North Carolina south to Florida and west to Texas.

ADDITIONAL NOTES: The seeds produced by this species are generally good food items for the Northern Bobwhite, songbirds, and small mammals. Dollarleaf is also utilized as a food plant for white-tailed deer.

Pencil Flower

Stylosanthes biflora

This inconspicuous native legume grows mixed in with the low-growing grasses in the longleaf ecosystems' herbaceous ground cover.

Pencil flower.

DESCRIPTION: The delicate wiry plant does not exceed 1½ feet in height and has small trifoliate compound leaves. The leaflets are bristle tipped, measure 1 inch long by ¼ inch wide, and have distinctive veins that run parallel to the leaf margin. The bright pencil yellow-orange flowers are seen throughout the summer.

HABITAT: It occurs in a variety of longleaf pine habitats, from sandhills to rolling hills with sunlight reaching the ground.

ADDITIONAL NOTES: The seeds of pencil flower are eaten by Northern Bobwhite and other birds. The plant is sometimes consumed by white-tailed deer and grazed by gopher tortoises.

Goat's Rue

Tephrosia virginiana

This common legume species grows across the range of longleaf pine. At one time this plant was fed to goats, as it was thought to improve milk production. However, due to the presence of toxic rotenone in the plant, this practice was soon discontinued, and the plant earned its common name. Another common name for this species is devil's shoestring, which refers to the plant's long, stringy roots.

Goat's rue flowers.

Goat's rue seed heads.

DESCRIPTION: This erect, perennial herb grows to a height of 1–3 feet and is covered in fine gray hairs. Its feather-like leaves can be as long as 5½ inches. The flowers are bicolored (pink and yellow/white) and appear in the spring or following a fire. The seed pod is a legume that measures up to 2 inches in length and holds several seeds.

HABITAT: It occurs throughout the eastern United States and can be found in a wide range of habitats. It is especially common in longleaf pine sandhills and rolling hills ecosystems.

ADDITIONAL NOTES: The seeds of this species are eaten by Northern Bobwhite and other seed-eating birds. The foliage is typically eaten by white-tailed deer and gopher tortoises, but only if few other foods are available. Because of the clumped, bushy structure of the plant, goat's rue also provides cover for ground-nesting birds or ground-dwelling reptiles. The seeds of this plant are commonly predated by weevils. Other common species include scurf hoary pea (*Tephrosia chrysophylla*), Florida hoary pea (*Tephrosia florida*), and spiked hoary pea (*Tephrosia spicata*).

St. Andrew's Cross

Hypericum hypericoides

DESCRIPTION: This delicate woody shrub is generally shorter than 1 foot but can reach 3 feet tall. The stems are dark brown and slender, with narrow leaves arranged oppositely on the stem. The leaves are small, measuring up to 1 inch long by ¼ inch wide. This summer-blooming plant has 4 lemon yellow petals arranged in a cross-like pattern with numerous yellow stamens emerging from the center of the flower. The flowers are supported by two

St Andrew's cross.

large green sepals (leafy structures that surround the flower) when flowering. These sepals will persist and enclose the developing capsule of seeds.

HABITAT: It grows in a variety of longleaf ecosystems, from dry sandhills to mesic uplands to wet flatwoods. It is wide ranging, occurring from New England to Florida and west to Texas.

ADDITIONAL NOTES: The seeds produced by St. Andrew's cross can sometimes be utilized by Northern Bobwhite and songbirds as a food source. Other common species include sandhill St. John's wort (*Hypericum tenuifolium*), pineland St. John's wort (*Hypericum suffruticosum*), and hairy St. John's wort (*Hypericum setosum*).

Savanna Meadow Beauty

Rhexia alifanus

This is a striking flower in longleaf pine forests. The bright pink flowers and urn-shaped seed pods atop a slender stem with appressed waxy leaves are distinctive and help identify this common wildflower species.

Savanna meadow beauty.

DESCRIPTION: This is a smooth, unbranched perennial plant with leaves arranged oppositely on the stem. It can reach a maximum height of 3 feet. The bright pink flowers are formed at the top of the plant during the summer. The fruit is a capsule that looks like a small pitcher or urn.

HABITAT: It grows in moist to wet seepage slopes and pine savannas. However, it can also be found in mesic longleaf pinelands.

ADDITIONAL NOTES: This plant is one of many species of *Rhexia* that occur in longleaf habitats; however, its plant family is predominately tropical in distribution. Some of the other species that can be found with longleaf include Maryland meadow beauty (*Rhexia mariana*), yellow meadow beauty (*Rhexia lutea*), and handsome Harry (*Rhexia virginica*).

Orange Milkwort

Polygala lutea

This is a common groundcover species in longleaf forests with wet soils. The bright orange and the overall shape of its flower arrangement has earned this species an alternate common name of bog Cheeto. It is usually found growing along with pitcher plants and other bog species. Another common name for this species is candyroot because of the licorice smell of the roots.

DESCRIPTION: The smooth stems of this herbaceous biennial plant species are generally less than 1 foot in height and grow out of a cluster of spatula-shaped basal leaves. The leaves are somewhat succulent to the touch. The branched stems each hold a cluster of bright orange flowers. This cylindrical cluster can be up to 1½ inches long and ½ inch wide. Orange milkwort blooms throughout the growing season.

HABITAT: It is found in the majority of the longleaf range and grows in wet savanna, flatwoods, and bog ecosystems.

Orange milkwort, also known as a bog Cheeto.

Pitcher Plants

Pitcher plants are one of several different kinds of carnivorous plants that can be found within longleaf pine forests. These plants evolved modified leaves to trap and digest arthropods to obtain nutrients that are not easily absorbed by roots in acidic, organic soils. Some species are quite rare. The overall shape and size of the pitchers indicate the type of prey they attract. The species described below are all adapted to attract arthropods.

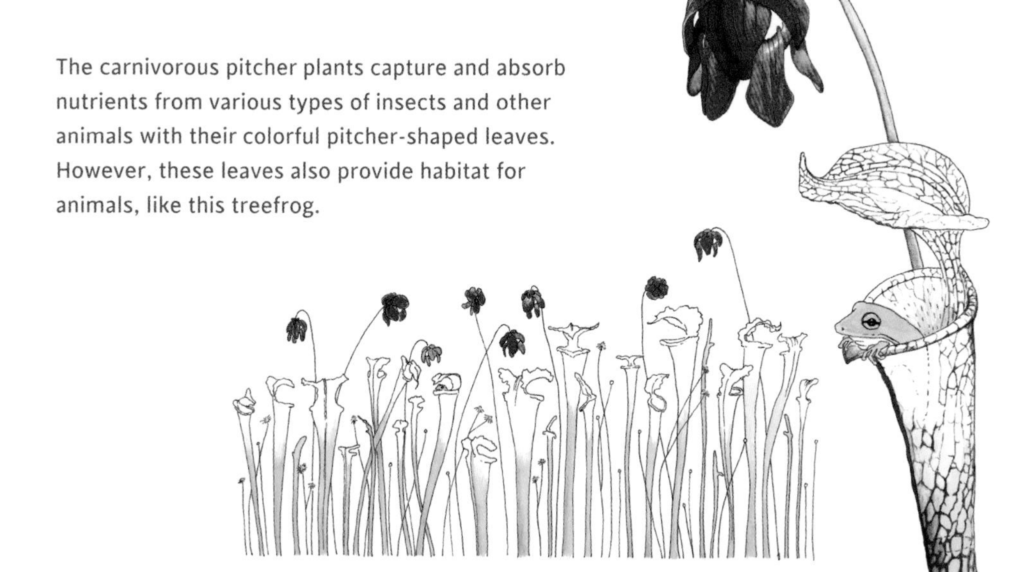

The carnivorous pitcher plants capture and absorb nutrients from various types of insects and other animals with their colorful pitcher-shaped leaves. However, these leaves also provide habitat for animals, like this treefrog.

Trumpet pitcher plant.

Trumpet Pitcher Plant (*Sarracenia flava*)

This has the widest range and is one of the more common species in the Southeast. It is also one of the tallest pitcher plants and lures arthropods with sweet nectar and sometimes bright coloration.

DESCRIPTION: Each plant of this herbaceous perennial can have many yellowish green leaves shaped like pitchers. Modified leaves are hollow and hold water. Arthropods are drawn to the coloration of the leaf as well as scent glands on the edge of the pitcher mouth. The unique nodding flowers are formed early in the spring and are bright yellow. Non-pitcher leaves are known as phyllodia.

HABITAT: This plant can be found in well-burned stream head margins, seepage slopes, and margins of lakes throughout the coastal plain.

Whitetop Pitcher Plant (*Sarracenia leucophylla*)

This is one of the showiest species of *Sarracenia* in the Southeast. The florist industry sometimes uses the pitchers for their bright colors and unusual shape in flower arrangements.

Whitetop pitcher plant.

DESCRIPTION: Each plant of this herbaceous perennial can have several tubular leaves (pitchers) that can be 37 inches tall. The lower part of the pitcher is green, but the top is white with red veins. This coloration makes it unique to the other species of pitcher plants in this region. The nodding flowers are formed from March to April and are dark red.

HABITAT: This can be found growing in bogs and wet savannas in the coastal plain of Georgia, Florida, Alabama, and Mississippi. The largest and healthiest populations of this species are in wet savanna habitats that are frequently burned.

Hooded Pitcher Plant (*Sarracenia minor*)

This species is shorter than the trumpet pitchers. Its name comes from the curved shape of the pitcher opening that resembles a hood.

DESCRIPTION: This herbaceous perennial plant has numerous 18-inch-tall pitchers that are mostly green with a red tint. Each pitcher also has white patches or "windows" on the back portion of the pitcher hood. These windows act as a false escape, depleting flying insects' energy as they repeatedly try to flee through a false hole. This spring-blooming species has yellow flowers that grow on stalks that are shorter than the pitchers.

HABITAT: It generally inhabits drier habitats than other species of *Sarracenia*. It can be found growing in wet pine savannas, flatwoods, and seepage slopes in Florida, Georgia, South Carolina, and North Carolina.

ADDITIONAL NOTES: Pitcher plants have limited pollinator value. However, they have been known to be hiding places for native spiders, such as the green lynx spider, or native frogs, which use the trumpet areas to ambush insects. Nonnative fire ants have been known to cut apart this plant in search of animals trapped inside.

Hooded pitcher plant.

Fixated on Nitrogen

When frequent fires pass through these forests, important nutrients like organic nitrogen may literally go up in smoke. Over time, loss of this important nutrient could lead to serious trouble for plant growth. As seen in this guide, longleaf pine forests are home to many species of native peas and beans. Legumes are widespread on the forest floor, with over 40 distinct species identified.

Legumes have formed a distinct relationship with bacteria found in the soil. These microscopic bacteria (called rhizobia) create growths on legume roots called nodules. These are often small but can be large, as is seen in agricultural peanuts. In exchange for providing energy to the nodules, the rhizobia transform nitrogen found in the air and eventually return it to the soil so it can be used by plants.

By contrast, plants found in soils that can be wet for extended periods (as seen in pine flatwoods or savannas) have adopted alternative strategies for limited nitrogen. In these areas, the acidic (pH <7) soils cause nitrogen to bond with organic compounds, making them unavailable for plant uptake. Carnivorous plants developed specialized trapping techniques to capture arthropods and absorb nutrients like nitrogen. Species of pitcher plants (*Sarracenia* spp.), sundews (*Drosera* spp.), Venus flytraps (*Dionaea muscipula*), bladderworts (*Utricularia* spp.), and butterworts (*Pinguicula* spp.) are all carnivorous plants found in longleaf pine forests.

Shrubs

Winged Sumac

Rhus copallinum

This is a common shrub seen in longleaf pine forests, especially areas recovering from large-scale soil disturbance like mechanical disking and logging. Though they can be problematic for some foresters trying to regenerate longleaf pine, they can provide benefits for various wildlife species. Aside from that, it can be an attractive shrub in the fall when its leaves change color.

DESCRIPTION: This fast-growing shrub/small tree can reach heights of 20–30 feet. In a typical longleaf pine forest, frequent fires keep these plants smaller. It has the ability to form stump sprouts and regrow from root fragments, serving it well following a fire. Its narrow leaves turn a brilliant reddish orange in fall and drop from the tree shortly thereafter. Dense clusters of small red fruits hang from the shrub and can last throughout the winter.

HABITAT: This is an early-pioneer species (one of the first to colonize an area) and is found frequently in rolling hills and mountain longleaf pine ecosystems.

ADDITIONAL NOTES: The drupes of winged sumac provide a valuable winter food source for a variety of birds. Likewise, rabbits have been known to eat its bark and twigs. The fallen leaves of the sumac are an important food source for the caterpillars of the red-banded hairstreak butterfly (*Calycopis cecrops*). Fragrant sumac (*Rhus aromatica*) and smooth sumac (*Rhus glabra*) are other species of sumac that occur in the Southeast.

Winged sumac.

Atlantic Poison Oak

Toxicodendron pubescens

This is closely related to eastern poison ivy and is similar in appearance except for its growth habit. Instead of a climbing shrub like poison ivy, this species grows upward and supports its own weight. Use the rhyme "leaves of three let it be" to help identify both species. As the name implies, poison-oak leaflets more closely resemble oak leaves than its local cousin.

DESCRIPTION: The low-growing shrub is usually under 3 feet tall. The stems grow from runners and create dense colonies. The glossy, green leaves that occur on the upper part of the stem are compound, with three lobed leaflets. This plant blooms in the summer, with drooping flower arrangements located beneath the leaves. Any portion of the plant contains urushiol, which can induce severe allergic skin reactions for sensitive individuals.

HABITAT: It occurs in dry, open pine forests.

Poison oak.

ADDITIONAL NOTES: Despite the adverse effects of this plant on people, it provides quite a bit of wildlife benefit. The fruits are eaten by a variety of bird species, and the leaves provide good forage for select mammal species. Other common species include poison ivy (*Toxicodendron radicans*) and poison sumac (*Toxicodendron vernix*).

Slimleaf Pawpaw

Asimina angustifolia

This small shrub occurs from South Georgia westward to Alabama and north Florida. Like many of the other pawpaw species, this plant emits a strong odor that some say is reminiscent of green bell peppers when the leaves are crushed. Pawpaws have one of the largest edible fruits in longleaf pine forests.

DESCRIPTION: The small shrub has one to few stems reaching a maximum height of 3–5 feet. Leaves are long and narrow, measuring up to 8 inches in length. In the spring it forms large yellowish white flowers in the leaf axils.

HABITAT: It can grow in dry to moist soils and occurs in longleaf pine flatwoods, longleaf pine sandhills, and rolling hills.

ADDITIONAL NOTES: Eastern Wild Turkey and a variety of mammals consume the fruit. Slimleaf pawpaw serves as an important pollinator plant in late winter and early spring and as a host plant for a variety of butterfly and moth larvae. Other common species include dwarf pawpaw (*Asimina pygmaea*) and bigflower pawpaw (*Asimina obovata*).

Slimleaf pawpaw leaves and flower.

Inkberry

Ilex glabra

This is an exceptionally flammable bush found in association with longleaf pine. Though inkberry is typically kept in check with frequent fires, the absence of fire can allow it to form dense thickets. Once fire does enter these thickets, it does so with great intensity.

DESCRIPTION: This native shrub stays green year-round. It is often small, growing 2–12 feet in height. Fire typically keeps the shrub smaller. It has dense, flammable foliage with leaves about 1½ inches in length. The leaves

Inkberry in flower.

Inkberry in fruit.

are smooth and green. The common name is derived from the small drupes, which were once used to make a commercial source of dark ink; it may also be called gallberry.

HABITAT: It is often found in wetter ecosystems, such as longleaf pine flatwoods or savannas. However, it can also be found in other longleaf pine ecosystems where water seeps from hill slopes.

ADDITIONAL NOTES: Its black fruit is an important food in the diet of mammals such as raccoons. Black bears have also been known to feast on its fruit. White-tailed deer may browse on inkberry twigs should little other food be available. Able to form thickets, inkberry can also be used as cover for small songbirds, such as Bachman's Sparrow. Of human interest, the nectar of inkberry is collected by bees and used to make a pleasant-tasting honey. Leaves were once used across much of the Southeast to make an herb tea. Other common species include Carolina holly (*Ilex ambigua*), large gallberry (*Ilex coriacea*), and American holly (*Ilex opaca*).

Yaupon Holly
Ilex vomitoria

This small tree or large shrub is common among coastal communities from North Carolina to eastern Texas. It may occur in a variety of ecosystems, including longleaf pine. Yaupon had a long history serving as a tea substitute. Longleaf pine forests with a heavy yaupon component are often suffering from a history of fire suppression.

DESCRIPTION: This shrub has small evergreen leaves and smooth gray bark. Flowers are hard to see but can have a strong, sweet fragrance. Yaupon plants are either male or female. Only female plants bear fruit. Female trees have bright red berries that ripen in the late summer or early fall.

HABITAT: This species is highly adaptable and able to survive in a wide range of conditions, including dry, sandy ridge tops and on the edge of wetland systems in both bright sun and dense shade.

ADDITIONAL NOTES: Berries are eaten by a variety of small mammals and songbirds. Flowers serve as an important nectar source for many pollinators. Thickets of yaupon can be important cover for many species of wildlife.

Yaupon holly leaves and fruit.

Saw palmetto resprouting following a prescribed fire.

Saw Palmetto

Serenoa repens

This shrub is a member of the palm family and has upright, fanlike leaves. It can form impenetrable thickets in some sites. Saw palmetto is common in flatwoods from Louisiana to extreme southeastern South Carolina.

DESCRIPTION: This species is a short palm with scaly, stout stems that grow horizontally along the ground. The leaves can reach up to 10 feet high. The leaf stalks have sawlike teeth along the edges. Saw palmetto flowers in the spring and produces numerous fleshy oval fruit.

HABITAT: This plant occurs on dry to seasonally wet pine sites.

ADDITIONAL NOTES: Saw palmetto is an inconsistent berry producer; however, when available, the fruits are used heavily by black bear, white-tailed deer, and feral hogs (*Sus scrofa*). This is an important plant for pollinators that produce honey. The berries are large, and when eaten by the black bear can pass through its gut relatively unharmed. In this sense, black bears are important in helping to redistribute palmetto seeds. Other common species include needle palm (*Rhapidophyllum hystrix*) and cabbage palm (*Sabal palmetto*).

Gopher Apple

Licania michauxii

DESCRIPTION: This low shrub reaches about 1½ feet tall, with leaves 1½–4 inches long and ½–1 inch wide. The leaves are mostly deciduous but can be evergreen in the southern parts of the range. The small white flowers appear at the top of the stem in early summer and the ¾- to 1-inch-long fruit ripens in early fall. The fruits are green at first and then turn a dirty-white color when ripe.

HABITAT: It can be found growing in sandhill habitat, roadsides, and coastal dunes in the southeastern coastal plain from South Carolina to Florida and west to Louisiana. This plant is often associated with turkey oak and longleaf pine.

ADDITIONAL NOTES: The fruits of gopher apple are often eaten by gopher tortoises. The flowers are also especially attractive to butterflies, making it a great pollinator plant.

Gopher apple leaves and fruit.

Titi
Cyrilla racemiflora

This small tree or large shrub is commonly observed in wetland communities associated with longleaf pine. It occurs throughout the range of longleaf pine in the Southeast. Many old-timers say that when the titi is in bloom fire activity will be high.

DESCRIPTION: Within the United States, titi is a small tree or large shrub with white-to-cream flowers born on clustered racemes (elongated cluster of flowers). It has smooth bark when young that sometimes ages to have flaky plates. Fruit are dry capsules and ripen in the late summer to early fall. Leaves are alternately arranged and oblanceolate (the pointed end near the base, wider toward the tip). In most of the region, titi is evergreen to semievergreen.

Titi.

HABITAT: It occurs in bottomland forests, wet pinelands, stream and river edges, swamp forests, and many other wet habitats. It is at least top-killed by fire, so heavy thickets of titi can indicate that fire has been excluded for some time.

ADDITIONAL NOTES: Seeds are eaten by a variety of songbirds and small mammals. Flowers serve as a nectar source for many native bees. However, titi is supposedly toxic to honeybees if they consume large quantities of its nectar.

Blueberries and Huckleberries

These shrubs are all members of the heath family. This group of plants provides tremendous wildlife benefit for a wide variety of species within the longleaf ecosystem. The soft fruit produced by these plants throughout the season are important components of animal nutrition. Most species rapidly resprout after being top-killed by fire. Fire is also known to increase fruit production.

Dwarf Huckleberry (*Gaylussacia dumosa*)

DESCRIPTION: This small shrub spreads from underground stems called rhizomes. The stems generally only grow up to 1½ feet tall. The leaves are small and glossy and have a pointed tip. The white flowers are small but showy and bell shaped. The berries are black and shiny. The plant will frequently resprout following fire.

HABITAT: This plant grows in dry to moist rolling hills, flatwoods, and savannas.

ADDITIONAL NOTES: The berries are eaten by various gamebirds, songbirds, and mammals. In some longleaf pine ecosystems, dwarf huckleberry serves as a very important source of nutrition for white-tailed deer. Other common species include dangleberry (*Gaylussacia frondosa*) and buckberry (*Gaylussacia ursina*).

Dwarf huckleberry.

Sparkleberry (*Vaccinium arboreum*)

This small tree or large shrub has a wide range throughout the Southeast. It is not a longleaf pine ecosystem endemic but occurs in dry, well-drained sites, so it can be a frequent component of some systems.

DESCRIPTION: They have gray smooth bark when young and red flaky bark at maturity. It has urn-shaped flowers that are born in early to mid-spring, with small black berries ripening in midsummer. The glossy leaves alternate on the stem. In part of its range, it is semievergreen.

Sparkleberry in flower.

HABITAT: It occurs in well drained, often infertile soils. In forests where fire is frequently used, sparkleberry tends to make small thickets where its habit can exclude fire. In addition, this species may occur in poor sites associated with longleaf pine such as oak scrub.

ADDITIONAL NOTES: Its flowers can be an important early nectar source for pollinators. The fruit of sparkleberry are eaten by a variety of birds, mammals, and insects.

Southern Highbush Blueberry (*Vaccinium formosum*)

Blueberries are mostly known for the large commercial fruits sold in grocery stores and roadside stands. Many of these commercial fruits owe their existence to the southern highbush blueberry, which occurs naturally in coastal communities from North Carolina to Florida into southern Alabama. This species is most abundant in the northern part of longleaf pine's range.

DESCRIPTION: This large shrub has urn-shaped flowers and dark blue berries. Flowering occurs in early spring and fruiting occurs early to midsummer. Leaf surfaces are smooth, but the leaf edges have small hairs. Leaves are arranged alternately along the stem.

HABITAT: It occurs in wet soils in the coastal plain, most often in longleaf pine flatwoods.

ADDITIONAL NOTES: The flowers can be an important early nectar source for many different pollinators. The fruit of southern highbush blueberries are eaten by a variety of birds, mammals, and insects.

Southern highbush blueberry in flower.

Deerberry (*Vaccinium stamineum*)

This small shrub of dry sites occurs throughout longleaf pine's range.

DESCRIPTION: Deerberry is a shrub with bell-shaped flowers and small green berries. Leaves grow alternately on the stem with smooth edges.

HABITAT: It can occur in a variety of ecosystems and has several varieties (forms that are often specific to habitat types) throughout its range. One variety, *arenicola*, occurs in pinelands from North Carolina to Southeast Georgia.

ADDITIONAL NOTES: The flowers can be an important early nectar source for many different pollinators. The berries are much larger, on average, than those of other similar species and are too large for most small birds to eat whole, as they do other blueberries. Instead, the fruits may be broken open and eaten, or they are eaten whole by larger animals, whereupon they scatter the seeds through feces. Also, unlike other blueberries, fruits fall when ripe, making them more accessible to ground-dwelling animals. The fruits are an important wildlife food in eastern North America. Many kinds of songbirds eat the berries, while white-tailed deer, in particular, eat the leaves, twigs, and berries. The berries are also relished by Northern Bobwhites, Eastern Wild Turkeys, foxes, raccoons, black bears, and small mammals. Other common species include Elliott's blueberry (*Vaccinium elliottii*), shiny blueberry (*Vaccinium myrsinites*), and dwarf deerberry (*Vaccinium stamineum* var. *arenicola*).

Deerberry in flower.

Allegheny Chinkapin

Castanea pumila

This unique-looking shrub may be mistaken as an American chestnut (*Castanea dentata*) sprout in some parts of its range. Chestnut blight (which wiped out the American chestnut tree) has caused Allegheny chinkapin to have localized declines, but otherwise it's fairly resistant to this fungus. In many areas of the longleaf pine range, it is an uncommon shrub.

DESCRIPTION: In many ways, this plant is a smaller version of the American chestnut tree. It is a moderate to large shrub with distinct leaves 3–6 inches long. The edges of the leaves have dull, toothlike edges. The top of the leaf is green, while the underside is typically fuzzy. Among the more distinguishing features are the chocolate brown to blackish brown nuts that are encased in spiny burs, with one seed per bur. When mature, the shrub can reach up to 30 feet tall. Fire typically results with chinkapin consistently being knocked back to shrub height.

HABITAT: Chinkapin grows well on sandy soils that are burned frequently. Though it does not tolerate most fires, it is able to easily resprout from the stump. It has the ability to spread by roots and form thick clusters of shrubs.

ADDITIONAL NOTES: Its nuts are among the more carbohydrate-rich hard mast in the longleaf pine forest. As such, they are excellent wildlife food and are consumed by a wide variety of birds and small mammals. Fox squirrels, Virginia opossums, Red-headed Woodpeckers, and Blue Jays are especially fond of chinkapin nuts. Squirrels may use the fallen leaves to make a wintertime home. White-tailed deer may browse the foliage in early spring.

Allegheny chinkapin.

Running Oak

Quercus elliottii

This shrubby oak can often be confused with young willow oaks (*Quercus phellos*). It has adapted well to the frequent fires in longleaf pine forests and provides valuable food sources for many wildlife species.

DESCRIPTION: This oak grows low to the ground and often is no more than 3 feet in height at maturity. The leaves are similar to willow oak but can be slightly rolled at the edges. The top of the leaf is lustrous green in the spring and summer, and the underside can be a fuzzy white. It has nearly black acorns, which mature in one year and are often available for wildlife food before any other oak species.

HABITAT: It is found on dry, sandy ridges and less dry rolling hill habitat. This oak is easily knocked back by frequent fires, yet will spread by underground stems. For this reason, it tends to be found in low-growing thickets.

Running oak acorn.

Running oak leaf.

ADDITIONAL NOTES: The oak produces abundant acorn crops, often when the plant is only a few feet tall. The acorns can be an early-season food source for white-tailed deer, Eastern Wild Turkey, black bear, and fox squirrels. Low-growing cover can provide refuge for many smaller animals such as birds, snakes, and small mammals.

Wax Myrtle

Morella cerifera

This is not only a common shrub in longleaf pine forests, it also serves a vital role in the ecosystem. Similar to legume species, the roots of wax myrtle have small, bulbous nodules that convert nitrogen found in the atmosphere and return it to the soil. This can be important when the soils run low on available nitrogen following frequent fires.

DESCRIPTION: Wax myrtle can be described as either a small tree or a large shrub. It can be found growing as tall as 20 feet high. Its leaves are a glossy green, up to 3 inches long and ¾ inch wide. These leaves are retained year-round. The plants are either male or female. Only female plants bear fruit. Female wax myrtle produces small flowers in the early spring that are replaced by grayish white to pale blue fruits in the summer. The fruit is heavily coated with wax and found in clusters on the stems.

HABITAT: This is adaptable to many different habitat types, though it is often found in rolling-hill longleaf pine habitats. As a thick, bushy shrub, wax myrtle can accumulate fallen pine needles of the longleaf pine. When fire moves through these suspended pine needles, it can cause the wax myrtle bush to go up in an explosive "whoosh" of flames. However, the roots of the wax myrtle typically survive most fires, and several stems can resprout from the root.

Wax myrtle.

ADDITIONAL NOTES: The waxy fruits of wax myrtle are an important food source for many birds, including Carolina Wren, Northern Bobwhite, and Eastern Wild Turkey. Likewise, being a thick shrub, many species of birds may use this shrub for refuge from predators. Wax myrtle has long been used as a source of wax in making bayberry candles.

American Beautyberry

Callicarpa americana

This is a common shrub in longleaf pine forests. Though it can be a nuisance for some foresters trying to restore longleaf pine, it does have wildlife value.

DESCRIPTION: This bushy shrub can reach 4–6 feet tall. The leaves fall off every year and regrow in the spring. Leaves are simple, averaging 5 inches long, and sit opposite one another on the twigs. The twigs are often arched, and it is not uncommon for the ends to be nipped off by browsing deer. Among the most distinguishing features on the beautyberry are the pink to red-purple drupes that are found in clusters on the branches.

American beautyberry.

HABITAT: It occurs frequently in rolling hill and mountain habitats. American beautyberry can be quite prolific following some disturbance that creates an opening in the forest. Killed back by fire, it easily resprouts or seeds germinate from the soil.

ADDITIONAL NOTES: The purple fruits are a choice food source for many birds, including Eastern Towhees, Eastern Bluebirds, Eastern Wild Turkeys, and Northern Bobwhites. Likewise, the fruits can be eaten by small mammals, such as the gray fox, Virginia opossums, and raccoons. The fruit can be an especially important food source in the early winter months. White-tailed deer will browse the twigs, especially in spring and summer when they are still succulent. Nonetheless, deer will turn to the twigs of beautyberry only when other primary food sources are scarce. Birds play an important role in dispersing the seeds of beautyberry and spreading it throughout the forest.

Woody Vines, Climbing Shrubs, and Brambles

Coral Honeysuckle
Lonicera sempervirens

This occurs throughout a wide range of habitats and can be found in some longleaf pine forests. The species does not seem to be favored by fire and therefore may be a good indicator of fire suppression.

Coral honeysuckle.

DESCRIPTION: It is a twining shrub (vine-like in habit) with glossy blue-green opposite leaves on the stem and conspicuous orange-red flowers. Flowers are trumpet shaped and occur in late spring to midsummer. Fruits are ripe in late summer to early fall.

HABITAT: Coral honeysuckle occurs in a variety of habitats from dry to moist soils in full sun to dense shade. It can be a component of longleaf pine forest edges, especially where fire is less frequent, such as in adjacent bottomland forest areas near streams and rivers.

ADDITIONAL NOTES: The flowers produce plentiful nectar and provide a food source for a variety of insects as well as the Ruby-throated Hummingbird (*Archilochus colubris*).

Carolina Jessamine

Gelsemium sempervirens

This woody vine is found in a variety of forest types. It is one of the first flowering plants seen each spring along roadsides across the region. Though native, it is not always a welcome plant in longleaf forests because of its tendency to form thick cover over native ground cover. An abundance of Carolina jessamine is a sign that proper fire management is not being applied to the site.

Carolina jessamine.

DESCRIPTION: This evergreen, woody vine climbs and trails. The dark green leaves are oppositely arranged on the stem and measure up to 2⅘ inches long and 1 inch wide. The flowers start appearing in March and are bright yellow and tubular.

HABITAT: This species can be found in a range of habitats including moist longleaf pine flatwoods, mountain longleaf pine, and rolling hills ecosystems. It can be found throughout the range of longleaf pine from Virginia to Florida and west to Texas.

ADDITIONAL NOTES: It provides an early season nectar source for hummingbirds and spicebush swallowtail butterflies. The leaves are sometimes eaten by white-tailed deer but contain toxic compounds for most mammals. Swamp jessamine *(Gelsemium rankinii)* is a wetland cousin that looks almost identical, but the flowers lack scent.

Purple Passionflower

Passiflora incarnata

This showy vine occurs across much of the Southeast. It occurs throughout longleaf pine's entire range. It is a member of a family of plants that are mostly tropical. Many kids from the South know this plant by the loud popping that the fruit makes when stepped on, giving it the nickname maypop.

DESCRIPTION: Flowers can be as large as 4 inches across with complex features. Flowers are generally purple and have a strong sweet fragrance. Flowers occur throughout the summer with each flower, opening only for a single day. Fruit ripens in late summer to early fall and is a large hollow berry with a membranous sac filled with sweet liquid surrounding the seed. The plant is a medium-sized vine with leaves resembling a trident.

HABITAT: Occurring in a variety of open habitats, it can be a weedy species occurring in disturbed soils and agricultural land. It occasionally occurs as a natural component of longleaf pine forests, especially where soil disturbance has taken place. It may also occur on roadsides, meadows, and power-line rights-of-way.

ADDITIONAL NOTES: Flowers are a nectar source for a variety of insects, especially nocturnal moths. Foliage is occasionally browsed by herbivores

Purple passionflower.

and is a larval host for the gulf fritillary butterfly. The fruit is eaten by mammals (including humans) and birds.

Dewberries

Southern Dewberry (*Rubus trivialis*)
Northern Dewberry (*Rubus flagellaris*)

The term "blackberry" typically describes a viny, thorned shrub or vine with edible fruits. Several different species can be called blackberries, but some are known as dewberry. Southern and northern dewberry occur throughout longleaf pine's range. Both occur in longleaf pine ecosystems and are similar in appearance and form.

DESCRIPTION: These dewberries are creeping brambles with leaves that each have 3–5 leaflets. Southern dewberry has hispid (hairy) stems in addition to the small thorns. Northern dewberry retains the small thorns but lacks hairs. Both species have white flowers in the spring followed by an aggregate of drupes (fleshy fruit containing one seed).

HABITAT: Both dewberries can occur in longleaf pine forests but may also occur in fields, power-line or roadside right of ways, disturbed land, agricultural borders, and generally anywhere else with plenty of sun exposure and well-draining soils.

ADDITIONAL NOTES: The flowers can be an important early nectar source for pollinators. The fruits of both dewberries are eaten by a variety of birds, mammals, and insects. Both species can form dense, low thickets that are vital cover for small- to medium-sized mammals.

Southern dewberry.

Northern dewberry.

Lance-leaf greenbrier.

Lance-Leaf Greenbrier
Smilax smallii

This evergreen semiwoody vine occurs throughout longleaf pine's range and can be a frequent component of moist sites in coastal plain pinelands.

DESCRIPTION: This species is a large vine that can climb to the canopy of small trees seeking sunlight. Alternately arranged leaves are extremely glossy because of their thick wax cuticle. Leaf margins are smooth and leaves are oval to lance shaped. Unlike most other smilax, stems are mostly free of prickles. Flowers are held in a spherical umbel (flowers on stalks arising from a central point). Berries are dull red when ripe.

HABITAT: It occurs in bottomland forests, roadsides, and pine flatwoods. Generally, it occurs in drier soils than the similar laurel leaf brier (*Smilax laurifolia*).

ADDITIONAL NOTES: The evergreen foliage of this plant can provide important cover, in the winter months especially, for a variety of animals. Flowers can serve as a nectar and pollen source for a variety of insects. Berries are eaten by birds, mammals, and insects. Herbivores may graze on young shoots, especially after a fire. Other common species include saw greenbrier (*Smilax bona-nox*) and laurel leaf brier.

Trees

Wild Persimmon

Diospyros virginiana

These are common trees found in longleaf pine forests and are prized by animals and humans alike for their fruit.

DESCRIPTION: This medium-sized tree is 30–60 feet tall. Leaves are ovate and about 5 inches long. The bark pattern on older trees is rough and resembles alligator skin. In the fall, glossy green leaves become tinged with yellow and eventually drop off. The profile of this tree when leafless is quite distinctive. The tree trunk is straight, but branches stick out at many crooked angles.

"Make a crust like you would any other pie crust and take your persimmons and wash them. Let them be good and ripe. Get the seed out of them. Don't cook them. Mash them and put cinnamon and spice in and butter. Sugar to taste. Then roll your dough and put it in custard pan, and then add the filling, then put a top crust on it, sprinkle a little sugar on top and bake."—Millie Evans, a formerly enslaved woman from Arkansas, giving her recipe for persimmon pie as part of a 1936 interview for the Federal Writers' Project.

Wild persimmon leaf.

Wild persimmon bark.

Wild persimmon fruit.

HABITAT: As these trees do not appear to handle the full force of fire well, they are often found along the edge of the forest, such as where it meets roads or fields, where fire is not as intense.

ADDITIONAL NOTES: The fruit is popular with Virginia opossums, raccoons, and gray foxes. Black bears are also fond of persimmons, and they have been known to climb trees and pull fruit from the branches. Because of the fondness Virginia opossums have for persimmons, they are sometimes called possum wood. The seeds of this tree are often spread in the scat of various animals.

Sourwood

Oxydendrum arboreum

This attractive tree is popular with some as a yard specimen. Largely thought of as a tree in hardwood forests, sourwood can also be found in longleaf pine forests growing in the mountains of North Georgia and Alabama.

DESCRIPTION: Throughout most of the year, its leaves are a lustrous green and seem to droop on the ends of the branches. The leaves are usually 4–8 inches long and oblong. However, the leaves are most noticed in the late

Sourwood leaf.

Sourwood bark.

Sourwood fruits.

summer and early fall, when they are among the first trees to turn a brilliant red. The trunk is usually crooked or leaning. Averaging 45–50 feet tall, the tree has white, bell-shaped flowers with fruits that can persist into the fall.

HABITAT: These trees are uncommon or altogether absent in longleaf pine forests over the majority of that tree's range. However, sourwood trees can be commonly seen growing at higher elevations on south-facing slopes along with longleaf pine in the mountain region. Their thin bark makes them susceptible to fire. Although these trees vigorously resprout following fire, frequently occurring fires will cause this tree to decline from the forest.

ADDITIONAL NOTES: Honeybees are attracted to the summer flowers of sourwood trees. In the mountains of North Alabama and Georgia, rural people have long used sourwood honey in curative concoctions. The canopy of larger sourwood trees will be used by gray squirrels to nest in the wintertime.

Oaks

Oaks belong to a group of trees that is found throughout many forests in the Southeast. They share common traits in that they typically have lobed, deciduous (lose every year) leaves. They bear acorns as fruit and often have wood that is strong. The common oaks found in the longleaf pine forests share further similar traits in that they are usually very tolerant of frequent fires.

Bluejack oak bark.

Bluejack Oak (*Quercus incana*)

This common tree is found in the drier areas in which longleaf pine grows. Like other scrub oaks, these are drab in appearance yet provide consistent benefits to wildlife because they are so hardy and produce regular acorns. Bluejack oaks are drought tolerant and have thick bark that protects them from most fires.

DESCRIPTION: This small tree reaches about 30–50 feet tall. The leaves appear bluish gray from a distance and might remain on the tree over the winter, making them conspicuous. The elliptical leaves grow to 3–5 inches long and about 1 inch at the widest point. One side is leathery green while the underside of the leaf has wooly fuzz. The bark is rough and blocky, providing fire protection. The acorns are ½–¾ inch long.

Bluejack oak acorn.

Bluejack oak leaf.

HABITAT: This scrubby oak is found on sandy ridges associated with longleaf pine.

ADDITIONAL NOTES: The thick crown of this tree can provide nesting areas for numerous songbirds such as Mourning Doves. Likewise, they can form thickets and provide suitable nesting areas for birds such as Eastern Towhee. When the trees die, cavity nesters such as Brown-headed Nuthatches will move in. Like other oaks, bluejack oaks will produce consistent acorn crops eaten by many animals. It is said that of the scrub oaks found with longleaf pine that fox squirrels prefer to consume bluejack oak acorns.

Turkey Oak (*Quercus laevis*)

These can often be found in great abundance in very dry longleaf pine forests, such as those found in the sandhills.

DESCRIPTION: The turkey oak does not reach vast heights, only growing to be 25–40 feet in most cases. The leaves resemble three lobes of a turkey's foot. Leaves are 4–10 inches long. Like other scrubby oaks, the bark is rough and thick, providing some protection from fire. Acorns reach about 1 inch long, and in some years large numbers of these acorns can be produced.

Turkey oak acorn.

Turkey oak bark.

Turkey oak leaf.

HABITAT: This tree is considered a scrubby oak found on dry barren ridges and sandy bluffs. The leaves do not burn well. In fact, it is not uncommon for thickets of turkey oaks to exclude fire from those areas of the forest. If the tree is killed by a fire, it can usually resprout from the ground.

ADDITIONAL NOTES: The acorns are nutritious and eaten by large and small mammals and some birds. Fox squirrels tend to bury the acorns, eating them only when not much else is available.

Sand Post Oak (*Quercus margaretta*)

This slow-growing oak has long been appreciated for its rot resistance and use as fence posts. The tree's widely spaced top has provided shade for many travelers through the "pine barrens."

DESCRIPTION: The three, big squarish lobes at the top of the post oak leaf form the shape of a 3- to 4-inch cross. The widest part of the cross is closer to the tip of the leaf. This medium-sized, scrubby oak can grow up to 65 feet tall. The acorn is about ¾ inch in length.

HABITAT: This species grows with longleaf pine on upland sites that have gravelly or sandy soils, such as ridges and sandhills.

Sand post oak leaf.

Sand post oak acorn.

Sand post oak bark.

ADDITIONAL NOTES: Acorns provide high-energy food during the lean fall and winter months. These acorns are considered important in the diet of many large and small mammals. The canopy and the leaves of the post oak are thick enough that birds, squirrels, and raccoons will build their nests in them. Mistletoe (*Phoradendron* sp.) can be found growing in the branches. Mistletoe berries provide food for passing songbirds. Though it would be a rare sighting in the longleaf pine range, this tree is also important in the life history of the spotted oak apple gall wasp (*Atrusca quercuscentricola*), which injects its egg on the post oak leaf. When the larvae hatch, a gall grows around them—the name derives from the gall's apple appearance.

Blackjack Oak (*Quercus marilandica*)

Seldom thought of as an attractive tree by foresters, these oaks are uniquely associated with poor growing conditions, such as those where longleaf pine thrives.

DESCRIPTION: This small oak tree grows in the mid-story of longleaf pine forests and has broad, triangular leaves 3–7 inches long. It only reaches approximately 50 feet tall. The leaves are coarse, leathery, and dark green on

Blackjack oak leaf.

Blackjack oak acorn.

Blackjack oak bark.

top; the underside is hairy and brown. The acorns are small, only measuring ¾ inch, and are half covered with a cap. The bark is coarse and thick, allowing this tree to be resistant to fire.

HABITAT: It is a common scrub oak on deep sandy ridges typically associated with longleaf pine. It typically is found growing in sandhill habitats; however, it can also be seen in rolling hills and sandstone ridges in the mountains.

ADDITIONAL NOTES: The acorns are small but are valued as a food source by most mammals, both large and small. Its limbs can be used for perches by birds such as Loggerhead Shrikes. Burned-out fire scars at the base of these trees may be used by reptiles such as the broadhead skink.

Sassafras

Sassafras albidum

This is a common small tree in longleaf pine forests. However, it can often be overlooked because fire typically knocks it back to a short height. When larger, the sassafras can be a favorite for many people because of the showy yellow leaves in the fall. It is also very aromatic, with both the crushed leaves and the roots smelling like root beer.

DESCRIPTION: Though sassafras can reach 60 feet tall, in the longleaf pine forest they are often seen as small saplings no taller than 10 feet. The leaves are 3–7 inches long, up to 4 inches wide, and come in three distinct shapes—elliptical, two lobed, and (rarely) three lobed. The bark on a mature tree is dark reddish brown and deeply furrowed. The bark of smaller trees is thin and yellow green.

HABITAT: It is often found in the mid-story or understory in longleaf pine forests. Along the forest edge, where fire can be less intense, the trees can grow much larger. Light fires typically kill back the small trees.

ADDITIONAL NOTES: During the summer and winter, sassafras twigs can be browsed by white-tailed deer. In some larger trees, fire may create deep scars that a variety of lizards, such as the broadhead skink, use. The green leaves are a larval food for spicebush swallowtail butterflies and are also eaten by rabbits and white-tailed deer. The drupe or fruit of sassafras can be eaten by a variety of birds, including Northern Bobwhites, Eastern Kingbirds, and

Sassafras bark.

Sassafras leaf.

Eastern Wild Turkeys. Humans have used this tree in the past for flavoring root beer and making tea. Safrole, however, an active compound found in the plant, is thought to be carcinogenic.

Wild Plum
Prunus umbellata

Some species of native plums have been found in the Southeast since the Ice Age, providing food to mastodons and other creatures. It has been speculated that the Native Americans in the Southeast cultivated this tree species and other varieties of native plums.

DESCRIPTION: This is a showy tree in the longleaf pine forest that makes up for its slight appearance with an abundance of small, white flowers. These trees are rounded off at the top and don't grow much taller than 20 feet. Likewise, branches spread out 15 feet or less. The leaves are tapered, serrated, and arranged alternately on the twigs. The underside of the leaf can look fuzzy white. The bark is thin, dark brown, and scaly.

HABITAT: This small tree can be found on dry rocky outcrops, such as those south-facing slopes where montane longleaf pine grows. Additionally, it grows in sandy upland forests. This tree has a thin bark that does not tolerate fire well. Where it forms thickets, fire has a hard time penetrating.

Wild plum leaf.

ADDITIONAL NOTES: The tart, purple fruit of this tree is eaten by many animals such as black bear, white-tailed deer, gray fox, and numerous songbirds. Additionally, the plums may form thickets in the absence of extensive fire, which provides excellent refuge for animals, such as Eastern Towhees, Northern Bobwhites, and rabbits. Eastern tent caterpillars (*Malacosoma americana*) will occasionally build nests (tents) in these trees.

Wild plum flower.

Wild plum bark.

Lichens

Dixie Reindeer Lichen and Powderpuff Lichen

Cladonia subtenuis and Cladonia evansii

These white, mosslike lichens occur in a variety of habitats throughout longleaf pine's range. Lichens are a symbiont of algae and fungi.

DESCRIPTION: These species are multibranched lichens that superficially appear mosslike by forming a mounded appearance. When dry, these species can be incredibly brittle; once they are moistened, they feel spongy to the touch. Reproduction is primarily asexual, with some spore production taking place by the fungal component. Reindeer lichen is generally flatter, with less rounded edges. Powderpuff lichen often appears cloud-like with billowy projections.

HABITAT: Both species are restricted to dry, bare, often infertile, exposed mineral soils. They can occur in abundance in sandhills but can also be found on remnant dune habitat near the coasts. Frequent fire may limit these species, though the fire itself doesn't spread very easily across areas where these lichens are dominant.

ADDITIONAL NOTES: The genus *Cladonia* can be a host to certain moth species. They may also be minor sources of food for herbivores, especially in the Southeast, where their importance is lessened by the availability of higher-quality food sources even in the winter. Other common species are reindeer lichen (*Cladina rangiferina*) and cup lichen (*Cladonia leporina*)

Dixie reindeer lichen.

Powderpuff lichen.

Reindeer lichen.

Cup lichen.

Nonnative and Invasive Species

As in many forested areas across the world, nonnative plants, animals, and even insects can wreak havoc if not adequately controlled. With few natural enemies, the invasive nonnatives can outcompete or prey on many native plants and animals. The following list includes some of the worst offenders that occur in longleaf habitats.

Cogongrass

Imperata cylindrica

This aggressive grass species from Asia was introduced as packing material and still used as an ornamental plant in the western United States. It can grow 2–3 feet tall, with grass blades that are about 1 inch wide with finely serrated leaf margins. The grass is largely not eaten by native animals, shades out other plants, and is difficult for smaller animals to move through. This grass burns very hot and is often the first grass to flower following a fire. For this reason, use of fire is not an effective control method.

Cogongrass seedhead.

Fire in cogongrass can be especially intense.

Japanese climbing fern.

Chinese tallow tree leaf.

Japanese Climbing Fern

Lygodium japonicum

A vine-type fern from Asia that is a relative newcomer to the Southeast, this is beginning to take over many longleaf pine forests. The deciduous leaves are lacy and delicate looking, but this plant is quite hardy. Infestation can become significant enough that it smothers out native vegetation. It has the ability to reach into the tree tops, allowing fire to travel from the ground to the tree canopy.

Chinese Tallow Tree

Triadica sebifera

This tree is said to have first been imported from China by Benjamin Franklin as an ornamental. It is especially problematic in wetter areas, like pine flatwoods and savannas. It is still planted in areas of the Southeast but is also expanding its range on its own. With thin bark, the tree is susceptible to fire. However, it also has the ability to colonize an area after a fire.

Feral Hogs

Sus scrofa

These can be up to 3 feet tall at the shoulder and 5 feet from snout to tail. Adults can weigh on average 180 pounds. Males (boars) are larger than females (sows). One favorite food of feral hogs is the soft root of longleaf pine seedlings. Hogs use their snouts to root up longleaf pine seedlings and consume the root system. Reports have been made of one hog being able to eat up to 400 longleaf pine seedlings per day. Their destruction doesn't end there, as they root up many areas, especially in pine flatwoods and savannas.

Woods hogs cause great damage to Louisiana's longleaf pine forests, as noted in 1951 by Walt Hopkins.

Red imported fire ants.

Distinctive mound of imported fire ants.

Red Imported Fire Ants and Black Imported Fire Ants

Solenopsis invicta* and *Solenopsis richteri

Both ant species were inadvertently introduced from South America through the Port of Mobile in the early 1900s. Color varies between black and red, and the two species have been known to hybridize. Ants are small, at ⅛ inch. Most distinctive, however, are their mounds (about 1 foot in diameter with no known center) and their aggressive behavior. These ants are known for their painful sting. They can also be fierce predators of baby gopher tortoises and birds (like Northern Bobwhite).

CHAPTER 4

Places to See Longleaf Pine

Despite a period of precipitous decline over the past century, there are still many areas of the South where mature, fire-maintained longleaf pine forests thrive. Some of these forests are found on private land, where access is restricted. Similarly, some of the best examples of longleaf pine forests are found on Department of Defense properties, which are generally inaccessible to the public. The following lists areas that are largely open to the public and include areas where you can see some of the best representations of longleaf pine forests. (For reference, 1 acre is approximately the size of one football field.) This is not an exhaustive list.

Virginia

It's uncertain how much of Virginia was covered with longleaf pine forests at the time of European settlement. A reasonable guess is that 1.5 million acres of southeastern Virginia was once covered with intact longleaf pine forests. A few scientists have estimated that longleaf pine reached all the way up the Tidewater area. Over the centuries, longleaf pine in Virginia has greatly declined. At one point in the recent past, Virginia was counting individual longleaf pine trees instead of acres of forest. The number of mature longleaf pine trees in Virginia is low, but acres planted in longleaf pine are increasing by the year. At the time of publication of this book, only one sizable area of mature longleaf is accessible to the public.

Sheep laurel (*Kalmia angustifolia*) in bloom at Blackwater Preserve, Virginia.

Blackwater Ecological Preserve. Isle of Wight County, Zuni area. This property owned by Old Dominion University has the distinction of the northernmost stand of intact, mature longleaf pine. The area is about 320 acres of mature longleaf pine with occasional loblolly pine and pond pine. It is mostly a longleaf pine flatwoods ecosystem. A number of cultural and natural resources can be seen, including several state-listed plants. The iconic pixie moss can be found here and nowhere else in Virginia.

From the village of Zuni, take State Route 614 South. At 1.6 miles, turn right to remain on State Route 614. Travel 2.8 miles to your destination on the right. The property is at 24326 Thomas Woods Trail, Zuni, Va. 23898 (36°48'44.01"N, 76°51'13.15"W). Permission to get onto the property must be granted in advance from Old Dominion University.

North Carolina

This state has a lengthy connection to its longleaf pine forests. For example, longleaf is woven into its state toast. The Order of the Longleaf Pine is one of the most prestigious awards that North Carolinians can receive. The state has long been connected to the naval-stores industry, which helped fund the establishment of towns like Wilmington and earned it the nickname of the Tarheel State. The Taft Arch in Wilmington welcomed the 27th president to the Land of Longleaf Pine in 1909. Towns in the central sandhill region of the state became worldwide destinations known for their restorative environments in the time of tuberculosis.

It is believed that at its peak in the 18th century approximately 12 million acres of longleaf pine forests existed in what is now North Carolina. One of the more recent surveys suggests about 400,000 acres remain. Much of what is found in North Carolina is on Department of Defense installations like Camp Lejeune and Fort Bragg (neither is open to the public). Longleaf can also be found on national forest property and a smattering of state parks and lands managed for public hunting and trapping (game lands). Remnants of old longleaf forests in towns like Southern Pines and Pinehurst have been integrated into old golf courses. In these areas, it's not unusual to see

From the Official State Toast of North Carolina

Here's to the land of the
long leaf pine,
The summer land where
the sun doth shine,
Where the weak grow
strong and the strong
grow great,
Here's to "Down Home,"
the Old North State! . . .

Taft Arch decorated for newly elected President William Howard Taft's 1909 visit to Wilmington, North Carolina.

Red-cockaded Woodpeckers and fox squirrels while playing a round of golf. The following are some locations where longleaf can still be found.

Weymouth Woods Sandhills Nature Preserve. Moore County, Southern Pines area. This preserve consists of about 900 acres total on three tracts—Weymouth Woods, Boyd Tract, and the Paint Hill Tract. All three are in close proximity to one another. The Boyd Tract, though heavily fire suppressed, is worth visiting due to the age of trees. One longleaf pine tree on the Boyd Tract is nearing 500 years old. The Weymouth Woods Tract has mature longleaf pine with wiregrass and other notable understory plants in areas.

One area in particular along the Pine Barrens Trail is at about 35°8'56.57"N, 79°21'53.01"W. To reach the area, navigate to the visitor center, which will have up-to-date information on where best to view longleaf pine. At the intersection of West Pennsylvania and Broad Street in the town of Southern Pines, head southeast on West Pennsylvania Avenue 0.2 mile. Turn right on South May Street and travel 0.3 mile. Turn left onto East Indiana Avenue and travel for 1.8 miles. Turn left onto South Fort Bragg Road, becoming North Fort Bragg Road, and travel for 0.6 mile. Turn left at the marked entrance of Weymouth Woods and follow the paved entrance road for 0.2 mile into the visitor center parking lot. The nature center is at 1024 Fort Bragg Road, Southern Pines, N.C. 28387.

Holly Shelter Game Lands. Pender County, Topsail area. Owned and managed by North Carolina Wildlife Resources Commission, these game lands cover 63,580 acres. Longleaf pine can be found throughout the area. The area is sandy soil among abundant low-lying swampland. It lacks the abundant scrub oaks found in other sandhills and is only about 50 feet above sea level. It also lacks the abundant waxy-leaved shrubs found in many flatwoods ecosystems. This area doesn't fit neatly into a description of a sandhill or flatwoods habitat, though is probably closer to the latter given that it lies near the coast.

One location has several hundred acres of mature, widely spaced longleaf pine growing over a carpet of wiregrass (near 34°24'37.02"N, 77°39'26.72"W). From the intersection of Jenkins Road / Country Club Road and Highway 17 in Topsail, travel northeast on Highway 17 for 2 miles. Turn left on Lodge Road and into a church parking area, also on the left, where you can either stop and walk in or, depending on the conditions, continue for 2.6 miles on Lodge Road. Lodge Road is a loose, sandy road in areas and under certain conditions is challenging to navigate.

Croatan National Forest. Carteret County, Broad Creek area. This national forest has 160,000 acres of pine forests, pocosins (shrubby swamps), and hardwood forests. There are many examples of longleaf pine forests scattered throughout.

Near 34°45'48.40"N, 76°57'22.30"W are several hundred acres of mature longleaf pine flatwoods. From Highway 24 in Broad Creek, turn onto and follow Nine Mile Road (County Road 1124) north/northwest 3.2 miles. Turn left on Millis Road (Highway 1112) and travel 0.5 mile, where the pavement ends, giving way to an unpaved surface. Continue on Millis Road 0.9 mile. Pull to the side, park here, and walk out into woods a couple thousand yards on the left (south) side of the road. Road conditions are accommodating to most traffic. The Ranger District Office is at Croatan Ranger District, 141 East Fisher Avenue, New Bern, N.C. 28560.

Carver's Creek State Park. Cumberland County, Carvers Creek area. This state park is home to thousands of acres of longleaf pine in sandhill habitat. Longleaf pine forests near the Sandhills Access are among the most picturesque.

The park office is at 2505 Long Valley Road, Spring Lake, N.C. 28390. From this office, take Long Valley Road south 0.3 mile. Turn left (northeast) onto Highway 1451 / East Manchester Road and travel 1.7 miles to Lillington Highway. Turn left and travel for 0.8 mile. Turn right on McCormick Bridge Road and travel for 3.2 miles. Turn left on Johnson Farm Road. At 6 miles, the road becomes Elliot Farm Road. Travel on this road for 1.7 miles. Turn right onto Elliot Bridge Road and travel for 0.5 mile. Turn right onto Highway 401 and travel 3.1 miles. Turn right onto McCloskey Road and continue 1.9 miles. The state park parking area is on left, near 35°10'11.39"N, 78°53'39.17"W. From the parking area, the 4-mile Longleaf Trail is a scenic hike through a longleaf pine forest.

Sandhills Game Land. Richmond County, Hoffman area. This large game land has about 60,000 acres and ample opportunities to see mature longleaf pine in sandhill habitat. One of the better locations is near 34°59'22.13"N, 79°30'30.24"W. At this location you'll see longleaf of multiple age classes, including mature longleaf pine growing above a carpet of wiregrass.

At the intersection of U.S. Highway 1 and School Drive in Hoffman, turn onto School Drive (southeast) and cross the railroad tracks. Follow School Drive all the way until it terminates on State Route 1602 / Butler Drive. Turn left on Butler Drive and follow for 1.0 mile, where it turns into State Route 1302 / Hoffman Road at this intersection. The road also turns to dirt in this area. Stay on Hoffman Drive for an additional 2.9 miles. Turn left (east) on Gardner Farm Road and follow for approximately 1.4 miles. Turn left (northwest) on Strausburg Lane and travel 0.9 mile. The stand is on the left (west), near 34°59'27.33"N, 79°30'23.53"W. The Sandhill Game Land Office is at 8 Hoffman Road, Hoffman, N.C. 28347. The office has additional information on longleaf worth seeing and conditions of the roads, closures, and so on. Roads can be deep sand and present difficult driving in areas.

Bladen Lakes State Forest. Bladen County, Colly Creek area. This state forest encompasses 159,885 acres of pine forests and embedded circular swamps called Carolina bays. The Addie Barnes Tract of the state forest contains approximately 160 acres of +90-year-old longleaf pine with a wiregrass carpet underneath. Red-cockaded Woodpeckers can be seen on occasion flying from tree to tree.

That tract is near 34°42'28"N, 78°28'30"W. From White Lake and the intersection of State Route 41 and U.S. Highway 701, travel north on Highway 701 for 4 miles. Though somewhat hard to see, Carter Blueberry Road is on the left. Turn left onto this road and travel 0.5 mile. Carter Blueberry Road is a primitive road and requires cautious driving be exercised. For more information, the forest supervisor's office is at 4470 Highway 242 North, Elizabethtown, N.C. 28337.

Uwharrie National Forest. Montgomery County, Daniel Mountain area. Though there are larger and better representative stands of mountain longleaf pine ecosystems in Georgia and Alabama, this stand is unique in its location. This approximately 60-acre area of longleaf pine is growing in mountain habitat in North Carolina. Near 35°25'4.82"N, 80° 2'0.47"W various age groups can be found, and some longleaf pine trees are 150 years old. Longleaf is growing on shallow, rocky soil up to an elevation of 875 feet above sea level.

At the intersection of State Route 109 and Reservation Road / State Route 1153, enter the Uwharrie National Forest. Follow Reservation Road for approximately 0.5 mile. Turn right onto Moccasin Creek Road / Forest Service 576 and continue just 0.2 mile. Turn left onto Cotton Place Road (dirt Forest Service Road 555) and travel 1.2 miles. Cotton Place ORV (off-road vehicle) trailhead parking is on the right. Park here and don't go up the ORV trail; instead, in the northwest corner of the parking lot is an unmarked fire line that is also used as a hiking and horse trail. Follow this fire line for about 1 mile to a saddle; the longleaf will start to come into view shortly before the saddle. Get to the saddle and go up the slope, where most pines are located. The ranger district office is at 1816a East Main Street, Albemarle, N.C. 28001, and can provide more information and information on local conditions.

Green Swamp Preserve. Brunswick County, Supply area. Owned by the Nature Conservancy, the property is open to the public from sunup to sundown. The majority of the 17,424-acre property is pocosin (shrub-dominated

Fall wildflowers in the longleaf savanna at
Green Swamp Preserve.

bogs), but longleaf pine savanna habitat can be found throughout. A trail marked by red diamond markers traverses through several of these impressive pine savannas. Many carnivorous plants and orchids can be seen on this preserve.

The parking lot is near 34°5'35.59"N, 78°17'57.31"W. From the intersection of State Highway 211 and U.S. Route 17, turn onto Highway 211 North / Green Swamp Road. Follow this road for approximately 5 miles. The parking area for the trailhead is on the right, indicated by a small parking area sign.

South Carolina

Longleaf pine can still be found in relative abundance in regional pockets in South Carolina. Many forests are found in the sandhills region, while others are found in the Lowcountry. A reasonable estimate of longleaf pine in South Carolina suggests that upward of 9.5 million acres was found at the time of European settlement. Maps in the late 1600s describe a "great savanna" outside of Charleston and likely described a vast grassland dotted with longleaf pine. Revolutionary troops skirmished in longleaf pine forests during the Battle of Camden. Georgetown and other coastal cities flourished from the exportation of longleaf pine timbers. Wealthy northern investors retreated to Aiken and elsewhere, establishing preserves like Hitchcock Woods that still harbor longleaf pine forests today. Ammunition-ignited fires at Fort Jackson kept longleaf pine forests hospitable for rare species, including Red-cockaded Woodpeckers. In 1989, Hurricane Hugo decimated Francis Marion National Forest, though the longleaf pine forests there have since recovered. Today, a reasonable guess is that less than 600,000 acres of longleaf pine forests remain in South Carolina. Planting efforts continue to increase that acreage, while the rapidly expanding human population further strains those older forests, which require frequent fire. Longleaf pine forests can be found on both public and private lands. Some of those areas include the following.

Carolina Sandhills National Wildlife Refuge. Chesterfield County, McBee area. There are many representative longleaf pine areas on this large, 45,348-acre refuge. Most fall under the category of sandhill ecosystems.

Among the higher quality area is compartment 18, cluster 13. The area contains seedling stage through old-growth stage longleaf pine. It has grassy understory and slight topography. Rare bird species such as Red-cockaded Woodpecker and Bachman's Sparrow can be found here. The main office of the refuge has posted hours but is a good location from which to start as they will have information on local conditions. They will also have species lists specific for the refuge.

To visit some of the higher-quality longleaf pine forests, travel 2.0 miles north on Wildlife Drive (workshop entrance) from the intersection of U.S. Highway 1. The intersection with Old Wire Road near 34°31'38"N, 80°13'37"W is a good place to pull over and walk out into the longleaf forest. The road system within the refuge is good. The refuge office is at 23734 Highway 1, McBee, S.C. 29101.

Lewis Ocean Bay Heritage Preserve. Horry County, Ocean Forest area. This 10,427-acre preserve near Myrtle Beach has had its share of challenges. Between hurricanes, wildfires, and a rapidly encroaching urban area, this preserve may not have the biggest or oldest longleaf pine forest in South Carolina. However, it makes up for this with its impressive groundcover plants. A number of interesting insectivorous (carnivorous) plants can be found, including the Venus fly trap and several pitcher plant species. There are also azaleas, orchids, and lilies on site. The area is also home to the largest black bear population in South Carolina. Red-cockaded Woodpeckers, Brown-headed Nuthatches, and several neo-tropical migratory warblers can be seen here.

From the intersection of International Drive and River Oaks Drive in Myrtle Beach, drive approximately 3.7 miles west on International Drive. Turn east (right) into a parking area / main entrance (near 33°48'21.08"N, 78°53'28.93"W) with an interpretive kiosk.

Francis Marion National Forest. Berkeley and Charleston Counties, Halfway Creek Trail and Awendaw Savanna areas. President Franklin Roosevelt created this 258,864-acre national forest in 1936. Though much of the land when acquired by the U.S. Forest Service was cutover timber land or worn-out farmland, today it contains many nice representations of longleaf pine forests. Hurricane Hugo in 1989 devastated much of these forests, so there are few old-growth longleaf pine trees on Francis Marion. But an aggressive prescribed-burning campaign had helped the longleaf pine forest recover. There are two must-see locations here: the Halfway Creek Trail and Awendaw Savanna Area.

The Halfway Creek Trail has a good access point and parking area just off Halfway Creek Road; it's probably the best and most accessible location to see longleaf pine forests up close. The parking turnout (33°2'18.22"N, 79°42'36.42"W) is about 2 miles south of the intersection of Steed Creek and Halfway Creek Road near Awendaw. Turn left onto Forest Service Road 5139 and park at the trailhead. Take the hiking trail south to a cypress pond. You can see Red-cockaded Woodpeckers and pitcher plants. The trail is several miles long and varies from flatwoods to savanna.

Awendaw Savanna is a high-quality example of a fire-maintained, wetland longleaf pine savanna with a high diversity of ground cover plants. Over 329 plant species have been documented here, including many orchids and carnivorous plants. From the intersection of Steed Creek Road and U.S. Highway 17 in Awendaw, drive east on Highway 17 for approximately 1.7 miles. Turn left (north), crossing over Highway 17 onto Duffield Road. Follow Duffield for 0.2 mile, then turn right (north) onto Forest Service Road 225. Parking is along this road (near 33°2'31.35"N, 79°35'36.53"W). For your first visit to Francis Marion National Forest, it is suggested that you visit the Sewee Visitors Center, 5821 U.S. Highway 17, Awendaw, S.C. 29429.

Hitchcock Woods. Aiken County, Aiken area. This property lies in the heart of Aiken. This 2,100-acre urban forest is privately owned and managed by the Hitchcock Woods Foundation. It is open to the public 7 days a week from sunrise to sundown. This sandhills region site supports a variety of habitats, including longleaf forests. Many old-growth longleaf pines grow on the property. One longleaf estimated to be between 300 and 400 years old is 105 feet tall and has a circumference of 140 inches. Coming back from a long history of fire suppression, an active prescribed-burning program (now 25 years in its application) is slowly attempting to restore this property's longleaf pine forest. Restoration of this property is ongoing, so if possible coordinate with the staff to determine the best location on the property to visit. Questions can be directed in person at the foundation office: Green House, 444 South Boundary Avenue SW, Aiken. Each of seven entrances to Hitchcock Woods takes you to a different part of the property. Hitchcock Woods is located near 33°32'57.96"N, 81°45'7.43"W.

Georgia

When James Oglethorpe carved out a colony in Georgia, he did so from an immense longleaf pine forest. Before Oglethorpe started felling trees, Georgia probably had as many as 23 million acres of longleaf pine, from the coast up to the mountains in the northwestern part of the state. The British were quick to establish a buckskin trade with the Native tribes already inhabiting the region. Hundreds of thousands of skins harvested from these seemingly endless forests started the journey from frontier villages like Columbus to Savannah and other coastal settlements for export.

During this time, settlers streamed through Augusta and other towns, grazing their cattle on wiregrass pasture on a tremendous sand ridge that stretched from Augusta to Columbus. The naval stores industry took root, and many towns like Waycross became a hub of activity. After the resins were extracted, the giant longleaf were cut and floated down the Altamaha and other rivers, supporting the growth of Darien, Brunswick, and other communities.

Following the Civil War, northern investors bought up large tracts of land in Thomasville and Albany to hunt quail. Today, although fewer than 1 million acres of longleaf can be found in Georgia, these two quail-hunting regions still support tremendous longleaf pine forests. Some of these areas near Thomasville have never been touched by an ax.

Along the sand ridge in Central Georgia, Fort Benning and Fort Gordon were among the Department of Defense installations established; farther east is Fort Stewart. These installations contain some of the most intact longleaf pine forests left in Georgia, though they are largely off limits to the public. Some of the locations open to the public include the following.

Ceylon Wildlife Management Area. Camden County, Billyville area. This property is owned and managed by the Georgia Department of Natural Resources. Longleaf pine in excess of 150 years old are found in a few stands here. Ground cover is a carpet of wiregrass, with shrub and plant species that are typical of longleaf pine growing in flatwoods and sandhill ecosystems. Gopher tortoises are found in great abundance in the deep sandhill areas; Ceylon has one of the largest gopher tortoise populations in Georgia. One stand worth seeing at Ceylon is off an unnamed road near 30°57'0.39"N, 81°37'21.14"W. It provides a good representation of what the forests would have looked like that greeted early European explorers to the region.

From Interstate 95 near Woodbine, take exit 14 (Woodbine exit). On Georgia Highway 25, travel west 0.2 mile. Turn left (south) onto Gap Swamp Road and travel for 2.0 miles. At the stop sign, turn left (east) onto Billyville Road. Continue 0.4 mile, at which point Billyville turns into Lang Road (it isn't marked as such). Follow on Lang Road for 0.75 mile, where asphalt gives way to gravel. Stay on this road for 1.2 miles. At this point, stay left. The road turns to well-packed dirt; continue 1.8 miles. Turn left (north) onto an unnamed woods road and follow for 1 mile. The stand is on the right. The dirt roads can be problematic during wet weather and require an all-wheel-drive vehicle.

Altama Plantation Wildlife Management Area. Glynn County, Altamaha area. This property is owned by the State of Georgia and managed by the Georgia Department of Natural Resources. Older specimen longleaf trees can be seen alongside the road from the main entrance to the Altama gate for over a mile. Many of these trees have catfaces (scars of past naval stores operations), faced away from the road. The stand is dense and linear, and the Georgia Department of Natural Resources is reintroducing fire to restore the ground cover. Parking at the gate and walking farther up the paved road affords more views of this stand.

Or, turn left at the gate and drive west along Ten Mile Road, crossing two distinct sandhills (both former barrier islands, geologically) separated by flatwoods and coastal wetlands. This route showcases other longleaf restoration work at Altama.

Remnant wiregrass and rare plants are found on the roadsides and stands where prescribed fire was reintroduced in 2018. Much of the property has 30-year-old longleaf and slash pine. Gopher tortoises and other wildlife are common throughout. This property falls within the Altamaha River Corridor conservation area, which stretches over 50 miles and 170,000 acres (at least 25,000 of which are being restored to longleaf pine–wiregrass) on both sides of the river.

In the Brunswick area, take exit 42 from Interstate 95, turning west onto Highway 99. Immediately on the north side of the road you will see signs for the Altama Wildlife Management Area; follow these signs.

Moody Forest Natural Area and Wildlife Management Area. Appling County, Baxley area. This 4,426-acre conservation area owned by the Nature Conservancy and Georgia Department of Natural Resources contains some of the last remaining biologically old-growth longleaf pine left in the state. Active restoration efforts have been taking place over the past few decades, and the rolling hill and pine flatwoods ecosystems are now in good condition. Among the better (and older) longleaf pine are near 31°55'0.79"N, 82°17'50.79"W.

From the city center of Baxley, follow U.S. Route 1 North for 8.3 miles. Turn right (east) onto Ashbury Church Road and continue 1.7 miles. Turn left (north) onto Spring Branch Road / Spring Branch Church Road and travel 0.8 mile. Turn right onto East River Road and travel 0.7 mile on dirt surface, which is drivable under most conditions. The parking lot is at the end of East River Road where it forks with Jake Moody Road. To the north is Tavia McCuean Trail, a 3-mile loop past some of the most scenic longleaf pine on the property.

Berry College. Floyd County, Lavender Mountain area. Though the entire Lavender Mountain once was mostly longleaf pine on its south-facing slopes, little of this forest remains today in good condition, due primarily to the exclusion of low-intensity fire. Approximately 170 acres on this mountain, however, have been actively restored for the past 20 years, now showing what old-growth longleaf pine looks like in mountain ecosystems at approximately 1,000-foot elevation.

Fire on private land in Thomas County, Georgia.

On the Mountain Campus of Berry College (near Rome, Ga.), travel west along the "Goat Road" to the Old Mill Wheel. Parking is available here. From the parking lot, find the 0.3-mile Longleaf Trail. The longleaf pine is along this steep trail near 34°19'38.15"N, 85°14'47.99"W.

Big Woods. Thomas County, Thomasville area. To view this remarkable stand of longleaf pine growing on private land, you can park on the shoulder of a public road. Otherwise, the Big Woods is private property and thus not open to public visitation: no trespassing. The Big Woods are old-growth forests growing in the rolling hills natural community type. The entire property is several thousand acres and historically has been managed as a Northern Bobwhite preserve.

From the intersection of West Thomasville Bypass / Highway 319 bypass west of Thomasville, take the Cairo Road northwest for 0.3 mile. Turn left (southwest) onto West Pine Boulevard and travel 0.25 mile to about 30°51'0.48"N, 84°0'50.64"W.

Florida

Though Florida has a long history of human habitation, it really wasn't until the end of the 19th century that significant forest timbering took place. For this reason, a fair number of historic photographs of longleaf pine in Florida provide a good idea of what this landscape once looked like. At one point, Florida had 17 million acres of longleaf pine forests scattered across mostly sandhills, flatwoods, and savannas. With the Spanish still in control of Florida, Pensacola and other port cities became important shipyard locations where straight longleaf pine was used for masts.

Many of Florida's historical cultures are connected to open piney woods with ample grass. Long before the human population exploded, the Pineywoods breed of cattle ranged freely, able to survive the harsh conditions of being left out in the woods. Cow hunters or cracker cowboys would round them up a few times a year to take to Tampa and elsewhere to be shipped out. Naval stores were also an important industry in Florida, and port cities from Jacksonville to St. Marks to Pensacola bustled with exportation activity.

The Olustee Experiment Forest in northeastern Florida would become an important center of research and development for experiments to maximize the production of pine "sap" (rosin, gum spirits). Tallahassee would draw northern investors to purchase land north of the city to set up places to hunt Northern Bobwhite and recreate. Investors also saw ample opportunity to liquidate the immense longleaf pine woods. Today, Millview and Shamrock are among the names of once-thriving, now-forgotten sawmill towns dotting Florida.

The Florida Natural Areas Inventory has confirmed that there are approximately 2.33 million acres of longleaf pine left in Florida today. One of the largest, most contiguous acreages is on Eglin Air Force Base (largely off-limits to the public). However, longleaf can also be found on all of the national forests in Florida, many wildlife refuges, and a number of state forests and parks. Some of the more notable areas to visit include the following locations.

Point Washington State Forest. Walton County, Blue Mountain area. Much of this part of Florida has seen significant alterations from decades of industrial forestry, urban expansion, and fire suppression. This was true of the Point Washington State Forest before restoration activities began. Owned and managed by Florida Forest Service, this 15,131-acre forest is still being restored through tree planting, burning, and clearing. A small, 30-acre parcel on the forest that escaped past logging contains old, flat-topped longleaf pine growing on deep sandy soils (near 30°20'46.40"N, 86°11'34.38"W). This area is not easy to reach.

From the intersection of Highways 98 and 331, travel east on Highway 98 for about 130 yards and turn right (south) onto a dirt road marked by Point Washington State Forest Public Entrance sign. This gate may be locked, in which case parking must occur here. If not locked, travel this sandy road for 0.9 mile. Turn right onto a smaller dirt road and continue 0.9 mile until it crosses another dirt road. Turn right onto this dirt road and travel 0.2 mile. From here, park and walk approximately 200 yards to reach the stand. Potentially poor road conditions recommend access with an all-wheel-drive vehicle. On occasion, the internal roads can be closed. The office headquarters is at 5865 East U.S. Highway 98, Santa Rosa Beach, Fla. 32459-6046.

Blackwater River State Forest. Santa Rosa County, Munson area. Owned and managed by Florida Forest Service, this state forest is worth a visit regardless of your intention: the campgrounds are top-notch for a state forest, with many places to swim as well as a handful of fishing lakes. There are thousands of acres of longleaf pine forests here. One of the better forests is near 30°51'41.60"N, 86°48'19.35"W. At this site, longleaf pine trees are mature and other younger age groups can be seen as well.

From Munson, travel on Highway 191 North to its intersection with Highway 4. Turn right (east) onto Highway 4 and travel 4 miles. Turn left (north) onto Old Martin Road and continue 0.7 mile. A good location to pull over is the intersection of Bungalow and Old Martin Roads.

Apalachicola National Forest. Leon County, Hilliardville area. Owned and managed by the U.S. Forest Service, this is the largest national forest in Florida at over 600,000 acres. It holds ample opportunities to see longleaf pine. That said, in 2018 Hurricane Michael decimated many stands on the western side of the forest, so the best locations to see longleaf pine are in Leon and Wakulla Counties. One area near 30°21'4.02"N, 84°30'49.38"W has mature longleaf pine with sandhills and flatwoods habitat in close proximity.

From the intersection of Springhill Road and Highway 267 North (Bloxham Cutoff Road) in Hilliardville, travel west on Highway 267 for 8.3 miles to Forest Service Road 305. Either side of the road has nice longleaf pine.

Ocala National Forest, Lake George Ranger District. Marion County, Prairie Pond area. Owned and managed by the U.S. Forest Service, these 430,000 acres have some of the best longleaf pine flatwoods and sandhill ecosystems in Central Florida. Aside from longleaf pine, there are open prairies, scrub, and numerous lakes and ponds. The most scenic longleaf pine is found near 29°23'52.07"N, 81°48'33.36"W.

From the intersection of Highway 315 and Highway 316 in Fort McCoy, travel east on Highway 316 for 9 miles. Turn left (north) onto Forest Service Road 88 and travel 2.3 miles. Roads at times can be soft sand, so a prior trip to the ranger district office at 17147 East State Road 40, Silver Springs, Fla. 34488, is recommended first.

Osceola National Forest. Baker County, Olustee area. This forest in northeastern Florida contains high-quality, mature longleaf pine growing in flatwoods ecosystems. Among the better stands are those found near 30°13'26.19"N, 82°24'30.41"W. This is also the location of the Macon Study. Established in 1958, this study site is one of the oldest ongoing prescribed-fire science experiments in the country.

From the intersection of County Road 231 and U.S. Highway 90 in Olustee, travel 1.2 miles east on Highway 90. Turn left (north) onto Highway 250A, traversing at-grade railroad tracks in the process. Travel 1.1 miles. Longleaf pine forest is on the right. Nearby is the Olustee Battlefield Historic Park, which also has a notable longleaf pine forest.

St. Marks National Wildlife Refuge. Wakulla County, 3 areas. This large, 68,000-acre refuge complex near Florida's Big Bend region is broken up into three units: St. Marks Unit, Panacea Unit, and Wakulla Unit. This large area presents opportunities to see mature, fire-maintained longleaf pine growing in many different natural community types. Many rare wildlife species can be seen, including the Red-cockaded Woodpecker, gopher tortoise, and numerous butterflies. A good time to visit the refuge is in the fall during the annual monarch butterfly festival. The visitor center has maps and species lists for the refuges. The visitor center is at 1255 Lighthouse Road, St. Marks, Fla. 32355.

St. Marks Wildlife Refuge, St. Marks Unit. Wakulla County, Newport area. Much of this surrounding region has been intensively managed for short-term fiber production. Bedding prior to tree planting was and still is a common practice to lift tree seedlings above standing water. However, bedding can be detrimental to wiregrass and other key understory plants. Near 30°9'54.92"N, 84°9'22.22"W is a small area on the St. Marks Unit containing some older longleaf pine trees (and slash pine) in a flatwoods setting. The understory has escaped bedding and has intact wiregrass, palmetto, and other species.

Near the town of St. Marks and at the intersection of U.S. Highway 98 and Lighthouse Road (County Road 59), turn south onto Lighthouse Road and travel southeast for 2.6 miles. Stop on the road shoulder on the left (east) side of the road.

St. Marks National Wildlife Refuge, Panacea Unit. Wakulla County. This area offers a unique transition of sandhills to flatwoods to open savanna. In the sandhills, stunted longleaf trees grow in company with numerous scrub oaks. Moving down gradient, the longleaf gets larger and the native grasses (like wiregrass) provide greater coverage. Slash pine begins to integrate with the longleaf as the soil becomes wetter.

From downtown Panacea at the intersection of Tully Avenue and U.S. Highway 98, travel north on Highway 98 for 1.4 miles. Turn right (east) onto Skipper Bay Road and park. A few hundred feet from this intersection is a sandy road that travels south/southeast and can be walked to a scenic location near 30°3'15.27"N, 84°23'10.16"W.

St. Marks National Wildlife Refuge, Wakulla Unit. Wakulla County. Some of the best longleaf pine on the Wakulla Unit is just west of Wakulla Field and north of Coggins Branch near 30°7'46.20"N, 84°18'18.00"W.

Near Shell Point, from the intersection of Spring Creek Highway and Shellpoint Road, is the Abe Trull Field Gate. Park here and hike along the trailhead (also called Vickers Drive) 0.1 mile. Take the Cross Seminole Trail and hike 0.8 mile. These are older longleaf pine flatwoods that grade into sandhill ecosystems. Wiregrass is mixed with palmetto, and on the sandier areas it would not be unusual to see a gopher tortoise.

Alabama

Longleaf pine can be found in a number of different habitat types in Alabama, from the 2,000-foot elevation in mountains near Talladega to their roots nearly touching saltwater on the coast. Alabama even had a historic stand (now gone) in the Cumberland region of North Alabama. A reasonable estimate is that the state once had 18 million acres of longleaf pine forests. Covering such a large area, longleaf pine forests played an important role in many industries and cultures in Alabama. Longleaf stands growing on mountain slopes were cleared and made into charcoal to fuel Birmingham's many furnaces. French settlers around Mobile would run cattle in the immense grassy areas growing under the longleaf pine forests there. Towns like Brewton, Tuscaloosa (Kaulton), Andalusia, and Sylacauga sprang up to harvest naval stores and seemingly endless longleaf lumber. Yale Forestry would take students to Alabama to learn about forestry. Places like the Solon Dixon Forestry Education Center would (and still do) train countless foresters on how to manage longleaf pine forests.

As Alabama's state tree, longleaf pine remains an important timber species, and several families still manage it specifically for products. There is slightly more than 900,000 acres of longleaf pine in the state despite an aggressive effort to replant it. It can still be found in relative abundance on state and federally owned property, much of which is open to the public.

Mountain Longleaf National Wildlife Refuge. Calhoun County, Anniston area. The 9,016-acre refuge was established in 2003 on the grounds of the old Fort McClellan. For decades, ordnances used by the military started range fires, leading to the maintenance of an exquisite mountain longleaf area. Some of these ordnances remain on the property, causing about two-thirds of the refuge to be closed to the public. Much of the longleaf is old growth or mature second growth. The most scenic old-growth longleaf is found on Caffey Hill and Red-Tail Ridge. Check in location is 407 Baby Bains Gap Road, Anniston, Ala. 36205; staff will provide guidance on where to go (and not go).

Talladega National Forest. This national forest is made up of three districts: Shoal Creek, Talladega, and Oakmulgee. All three have abundant mountain longleaf pine and longleaf pine found in the rolling hills habitat type. Prescribed fire is regularly used on much of this national forest. A few areas are considered old growth. No check-in is required, but the districts' respective ranger stations can provide the most current information.

Oakmulgee District of Talladega National Forest. Bibb County, Highfield area. The Highfield is approximately 4,000 acres of high-quality old longleaf in the rolling hills of Central Alabama. At the intersection of Forest Service Roads 738 and 726 near Keeton Corner, travel west on Forest Service Road 738 for approximately 1.4 miles. Turn south on Forest Service Road 723 and continue about 1 mile (near 32°57'31.27"N, 87°23'59.85"W). The road is passable in some SUVs. These roads are unimproved and isolated, and drivers should make appropriate contingencies. The Oakmulgee Ranger District Office is at 9901 Highway 5, Brent, Ala. 35034.

Shoal Creek District of Talladega National Forest. Cleburne County, Cole Cemetery and Rattlesnake Mountain areas. The Rattlesnake Mountain area contains thousands of acres of mature mountain longleaf pine. Given the elevation, longleaf is often found growing in close proximity to shortleaf pine. From Edwardsville, follow Highway 61 and signs in the direction of Coleman Lake. There are well-represented longleaf and shortleaf pine forests along this route. Highway 61 terminates and turns into the graveled Forest Service Road 500 (Skyway Road). Nice representations of mountain longleaf appear all along Rattlesnake Mountain and Skyway Road. There are a few old (flat-topped) longleaf here of unknown age. Follow this road 3.7 miles. Some of the nicest trees can be found near 33°44'9.47"N, 85°35'59.39"W, at 1,200-foot elevation. Most of the main road is passable by passenger car, depending on conditions. The district office is at 45 Alabama Highway 281, Heflin, Ala. 36264.

Talladega Ranger District of Talladega National Forest. Talladega County, Horn Mountain area. Much of the several thousand acres of old-growth longleaf pine growing in mountain ecosystems here are fire suppressed. At this elevation, longleaf begins to integrate with shortleaf pine, though there are many areas along the drive where only longleaf pine is found in the tree canopy. Longleaf is predominantly found on southerly facing slopes.

From Talladega town center, take Ashland Highway (Highway 77) south for 10.8 miles. Turn west on Forest Service Road 307, which is rough and steep. Follow this uphill for 3.5 miles to the microwave/radio tower: park here. There is old longleaf pine most of the way up the drive. Pinhoti Trail passes through this area, and if you walk downhill about 0.2 mile you'll see old-growth mountain longleaf that grade into shortleaf pine forests farther uphill (near 33°18'48.03"N, 86°4'2.36"W). Neither the drive nor the hike is easy, so be prepared. Either an all-wheel-drive vehicle with adequate clearance or a truck is required to reach Horn Mountain. The district office is at 1001 North Street East, Talladega, Ala. 35160.

Conecuh National Forest. Covington County, Boggy Creek area. Fire-managed longleaf pine stands can be seen in many areas here. Much of the Conecuh is rolling hill ecosystems, but there are also areas of flatwoods community types, some of which contain pitcher plants. Some of the best representations of longleaf pine forests are found in compartments 54 and 58 near the Boggy Creek area. Longleaf pine trees in this area are nearing a century old, with many Red-cockaded Woodpeckers using trees here.

At the intersection of Highway 4 and 137 near Wing, Alabama, turn west onto Highway 4. Continue 1.4 miles and turn north onto Forest Service Road 321 and park (near 31°1'32.62"N, 86°38'16.87"W). Under most conditions, a sedan-type vehicle is all that is needed to travel these roads. The district office is at 24481 Alabama Highway 55, Andalusia, Ala. 36420.

Tuskegee National Forest. Macon County, Alliance area. Even in this, the smallest national forest in the country, there are still ample opportunities to view longleaf pine growing in rolling hill habitat. In some areas, longleaf pine trees are approaching 100 years old and routinely subjected to prescribed fire. One of the better maintained stands is in the Alliance area.

Near the Notasulga area, exit Interstate 85 at exit 42 (Wire Road). Traveling Highway 80 East toward Columbus, after 3.6 miles, exit onto Highway 29 toward Auburn (northeast). After 1.5 miles, turn north (left) onto County Road 54. Mature longleaf pine will be found on both sides of this dirt road. The best stand is 0.6 mile down County Road 54 (near 32°29'40.68"N, 85°33'51.80"W). The district office is at 125 National Forest Road 949, Tuskegee, Ala. 36083.

Mobile Botanical Gardens. Mobile County, Mobile area. This is a 35-acre mature longleaf pine forest in rolling hill ecosystems (with some character of flatwoods community types). Staff have successfully restored the forest that greeted early French settlers to the region in the 1700s. Mature, fire-maintained longleaf abounds, as do many wildflowers and shrubs. Gopher tortoises can be found in this area. An interpretive kiosk is at the trail entrance. The longleaf forest area is open from dawn until dusk.

In Mobile, take exit 5A from Interstate 65 onto Springhill Avenue. Travel Springhill Avenue West for 1.6 miles. Turn left onto Pfc John O. New Drive. After a few hundred feet, turn onto Museum Drive. Continue on this asphalt road for 0.5 mile to the parking area. The forest is on the south side of the road (near 30°42'12.83"N, 88°9'38.17"W). The gardens are at 5151 Museum Drive, Mobile, Ala. 36608.

Wind Creek State Park. Tallapoosa County, Jackson Gap area. Much of the many hundreds of acres of mature longleaf pine here are fire suppressed, but progress is being made toward its restoration. A few smaller stands near the park entrance have been restored and have hiking trails throughout them. Part of the Central Pine Hills of Alabama, this area is distinctively longleaf pine in rolling hills ecosystems.

From Dadeville, follow U.S. Highway 280 West 10.2 miles. Turn left (south) onto County Road 21 (Coven Abbett Highway) and continue 4.2 miles. The park entrance is on Hodnett Drive. The longleaf pine forest is on the west side of County Road 21 (near 32°51'35.13"N, 85°56'11.68"W). This asphalt road has parking near the park entrance. The park is at 4325 Alabama Highway 128, Alexander City, Ala. 35010.

Mississippi

Much of the longleaf pine found in Mississippi was in the southern part of the state. Historically, 9 million acres of longleaf pine covered this state. The history of this forest in Mississippi is largely a utilitarian one. Huge trees received the attention of investors from the northern part of the Low Country, where timber supplies were already exhausted. This gave rise to many sawmills and logging towns. With the theme being to cut out and then get out, many of those once-thriving towns, like Brookhaven, Bowie, Griffin, Petal, and Rosine, are gone or are slowly fading away. Once cleared of timber, even the resinous stumps had value and were shipped to hubs like the (now long mothballed) Hercules Plant in Hattiesburg.

Following this period, a sizable portion of the land once covered with longleaf pine was purchased by the U.S. Forest Service. An army of tree planters with the New Deal's Civilian Conservation Corps planted longleaf pine in places. This gave rise to what we see today on DeSoto National Forest or Homochitto National Forest. Some longleaf forests are found on Camp

Shelby, though it is largely inaccessible to the public. Some open savannas were protected along Mississippi's Gulf Coast. Today, though fewer than 350,000 acres of longleaf forests are found in Mississippi, some areas remain, including the following.

Homochitto National Forest. Franklin County, Richardson Creek area. Though much of this is loblolly pine forest, there are longleaf pine in some areas. Longleaf can be found here mostly following high, dry ridgelines. One of the better locations with longleaf pines that are 80–100 years old is near 31°24'18.00"N, 91°1'55.20"W.

From Roxie at the intersection of U.S. Highway 84 and 3rd Street, take Highway 84 east for approximately 4 miles. Turn right (south) onto old Highway 84 and travel 1.4 miles. After a slight bend in the roadway, this road becomes Knoxville Road. Follow for 0.4 mile, then turn left onto Chapel Road. Travel 5.1 miles, then turn left onto Butler Graveyard Road. At 0.4 mile, longleaf follows the ridgeline here. Once you get off of Highway 82, the roads are mostly county maintained and can present questionable driving in heavy rain. Check with the Homochitto Ranger District at 1200 Highway 184, East Meadville, Miss. 39653, for other locations.

Pale pitcher plants (*Sarracenia alata*) at Buttercup Flats, DeSoto National Forest, Mississippi.

DeSoto National Forest. Owned and managed by the U.S. Forest Service, this area is in southeastern Mississippi. This national forest's 518,587 acres are divided into two ranger districts, the DeSoto and Chickasawhay. A number of different ecosystem types can be found, from savannas to flatwoods, sand-hills, and rolling hills. One particularly renowned pitcher plant savanna area is called Buttercup Flats, a +100-acre expanse of pitcher plants (primarily the pale pitcher plant, *Sarracenia alata*). Though old longleaf pine veterans exist here and there, when the U.S. Forest Service acquired the land, it was largely cutover. This forest is home to black pine snakes and federally threatened gopher tortoises. The predominant understory grasses that carry fire are several bluestem species. Nonnative and aggressive cogongrass requires aggressive intervention by the U.S. Forest Service.

DeSoto Ranger District of the DeSoto National Forest. Perry County, Beaver Creek area. In 2005, this forest district was impacted severely by Hurricane

Katrina. Though areas that had longleaf pine fared better than those with other pine species, this powerful storm left few areas untouched. Fast-forward +15 years, salvage operations, and multiple prescribed burns, the scars of the storm are harder to detect. At the intersection of Forest Service Roads 309 and 405 (Mars Hill Road), near 30°58'25.32"N, 88°54'15.84"W, are several hundred acres of mature longleaf pine. Openings created by Hurricane Katrina have filled in with longleaf pine saplings and seedlings, while frequent fires have since removed much of the storm debris. There is an active Red-cockaded Woodpecker colony nearby on Forest Service Road 405. Gopher tortoises and black pine snakes can also be seen in the area. The Desoto Ranger District Office is at 654 West Frontage Road, Wiggins, Miss. 39577.

Chickasawhay Ranger District of the DeSoto National Forest. Wayne and Jones Counties, Little Thompson Creek area. Several hundred acres of mature longleaf pine are found in this area. Red-cockaded Woodpeckers and gopher tortoises are frequent occurrences. Many thousands of acres of longleaf pine forest can be found en route to this stand on Forest Service Road 221A, and many of them are photogenic. This particular stand fits within the rolling hills ecosystem type, though pockets of sandhills can be found on some of the ridges (e.g., near 31°28'55.88"N, 88°56'52.26"W).

From Richton at the intersection of County Roads 15 and 42, travel north on County Road 15 approximately 8.2 miles. Turn right (east) onto Brown Cemetery Road and travel this hard-packed clay and gravel road for 1.4 miles. Turn onto Forest Service Road 205 and travel 2.6 miles. Road conditions on this forest service road will vary with weather conditions. When dry, a two-wheel-drive car can ride the road. When wet, clay-based roads will be slick. The ranger district office is at 968 Highway 15, South Laurel, Miss. 39443.

Sandhill Crane Wildlife Refuge. Jackson County, Fort Bayou Creek area. The 19,300-acre refuge consists of three land units on both sides of Interstate 10. Within its boundaries are found some of the last remaining, frequently burned longleaf pine savannas left on the Gulf Coast. In several areas the trees are very sparse, and some areas have old longleaf pine, slash pine, or a mixture of these two. One area with a scenic longleaf pine savanna is near 30°28'50.63"N, 88°47'15.43"W. The interior roads to reach this location are in good shape, but you must first secure a special-use permit from the refuge center to enter. The center is at 7200 Crane Lane, Gautier, Miss. 39553.

Grand Bay National Wildlife Refuge. Jackson County, Pecan area. This 32,000-acre refuge is in adjacent Mobile County, Ala. It also shares boundaries with the Grand Bay National Estuarine Research Reserve. Though much of the historic pine savanna is overgrown and fire suppressed in this general region, approximately 150 acres of fire-maintained savanna can be found here. Longleaf pine and some slash pine are widely scattered. The longleaf pine trees are old with gnarled flattened tops but are otherwise small, at 12 inches in diameter and 45 feet tall. Invasive cogongrass and feral hogs are a concern in this region.

Near the Mississippi–Alabama state line, take exit 75 on Interstate 10 (Franklin Creek Road exit). Turn left and travel south to Highway 90. Cross onto Pecan Road. At approximately 1 mile you'll cross railroad tracks. The road becomes Bayou Heron Road. About 0.7 mile down this road, the savanna will be on the right (west), near 30°26'3.24"N, 88°25'39.71"W. Travel 0.2 mile to the visitor's center, which can supply instructions on how to access some of these savanna areas. The visitor center is at 6005 Bayou Heron Road, Moss Point, Miss. 39562.

Louisiana

Growing on the north shore of Lake Pontchartrain once stood the most impressive forest of longleaf pine ever seen. It contained longleaf pine trees over 100 feet tall and wide enough that two adults could barely join arms around them. At Bogalusa these trees fed the Great Southern Lumber Company sawmill, at one time the largest sawmill in the world. However, Bogalusa suffered the same fate as other sawmill towns in the longleaf pine range when the forest was slashed. Other long-forgotten sawmill towns included ones named Longleaf and Tioga. Louisiana once had about 7.5 million acres of longleaf pine forest. *All* of it was cut. Some acres were reforested, but fewer than 500,000 acres of longleaf are in the state today. Gone is the immense forest of the north shore, never to be seen again in our lifetime.

What does remain is found on Fort Polk (largely inaccessible to the public), Kisatchie National Forest, and a smattering of other areas.

Alexander State Forest. Rapides Parish, Woodworth area. This historic forest is just under 8,000 acres. It is primarily managed for loblolly pine, though some older stands of longleaf pine can be found. Its 90 acres of longleaf pine (near 31°7'20.11"N, 92°29'10.85"W) are estimated to be 100–120 years in age. The landform is best described as rolling hills.

From Woodworth and the intersection of U.S. Highway 165 and State Highway 3265, travel on State Highway 3625 east for 0.4 mile to Indian Creek Road. Turn right (south) on Indian Creek Road and travel 1.2 miles. A sign for the Recreation Area is on the left (east). Turn left (east) onto Campground Road and travel on this road about 0.5 mile. The longleaf stand begins on the left (north).

Kisatchie National Forest. The Kisatchie is the only national forest in Louisiana. However, at 604,000 acres and with five ranger districts, it covers many different areas of the state. These districts include Calcasieu, Caney, Catahoula, Kisatchie, and Winn. Four of the five districts have natural longleaf pine forest (only the Caney District does not). Some of these areas are home to some of the best remaining longleaf in the western range of the species. In 1935, Congress created the Palustris Experimental Forest in the Evangeline Unit (part of the Calcasieu Ranger District) as a field research laboratory and demonstration site for new forestry practices. The study of longleaf pine is of particular importance for this experimental forest. The 7,500 acres within this experimental forest have been used for almost 60 years to monitor prescribed burning, tree spacing, tree growth, and so on.

Vernon Unit, Calcasieu Ranger District of Kisatchie National Forest. Vernon Parish, Longleaf Scenic area. This unit encompassing 85,000 acres has long been a showpiece of longleaf pine, equaling anything else found across the tree's range. Frequent fire has also promoted a rich understory of bluestems and three-awn grasses, forbs, and fall-flowering species. Longleaf specialists such as Red-cockaded Woodpecker, Northern Bobwhite, and the Louisiana pine snake can be found here. A few notable hillside seepage bogs are also worth seeing. There have been over 100 plant species documented in Cooter's

Bog (367 acres), Drakes Creek Bog (205 acres), and Leo's Bog (143 acres), including pale pitcher plants and other insectivorous plants. Other fire-dependent species can be found in these bogs, such as toothache grass and various bog buttons. Despite significant impacts by Hurricanes Laura and Delta in 2020, the Longleaf Scenic Area (Forest Service Compartments 10 and 14) remains one of the best locations to see longleaf pine (near 30°59'52.09"N, 93°8'7.07"W).

From Pitkin at the intersection of Louisiana Highway 113 and the Pitkin Highway (La. 10), follow Louisiana Highway 10 (Pitkin Highway) northwest/west for approximately 14.4 miles. Turn right (northeast) onto Forest Service

Road 400 (at a recreation area sign). Follow this improved road for about 3 miles to the Longleaf Scenic Area. There will be ample opportunities to see longleaf along the entire route. Forest Service Road 400 will fork at about 0.4 mile in. Stay right to remain on Forest Service Road 400 and continue 1.9 miles from the end of the asphalt. Turn left (north) onto Forest Service Road 444. Travel about 1 mile. The scenic area is on the left (west) side of the road. The road turns to dirt at some point but is passable under most conditions. If visiting, check the status of roads and trails, as they may still be closed due to hurricane recovery efforts. The Calcasieu Ranger District Office is at 9912 Highway 28, West Boyce, La. 71409.

Longleaf pine as far as the eye can see on the Vernon Unit of Kisatchie National Forest in Louisiana.

Kisatchie Ranger District of the Kisatchie National Forest. Natchitoches Parish, Longleaf Vista and Turpentine Hill area. Southwest of Natchitoches, Louisiana, this approximately 102,000-acre district provides ample opportunities to see longleaf pine forests. Among the more unique are the sandstone bluffs of the Longleaf Vista Area. Rugged by Louisiana standards, this area is an interesting look at longleaf pine growing in extreme rolling hills (almost montane) forest conditions.

Near Boyce, take exit 103 on Interstate 49. Turn west onto Louisiana Highway 8 toward Flatwoods and continue 9.6 miles. Turn right (north/northwest) onto Highway 119 and travel 8.2 miles. Turn left (west) onto Parish Road 830. A Forest Service sign here directs traffic toward the Longleaf Scenic Trail. Follow the Longleaf Scenic Road for nearly 3 miles, then turn right (east) up a small hill to a parking lot (compartments 45 and 46). Trails and vistas of longleaf pine forests can be found from here (near 31°28'30.93"N, 92°59'56.60"W). The Middle Branch Bog Natural Area (75 acres) and North Bayou L'Ivrogne Bog (85 acres) also provide opportunities to see longleaf pine in this district. Here the forest is growing in wet pine flatwoods and savannas. In these wet areas, longleaf pine is found with bluestems, meadow beauties, pale pitcher plants, and other insectivorous plants common in bogs and wet areas. Check with the local Kisatchie Ranger District at 229 Dogwood Park Road, Provencal, La. 71468, for road conditions and directions to reach some of the natural areas.

Texas

The piney woods of Texas conjures the image of an immense forest since, after all, everything is bigger in Texas. At one time over 500 sawmills were operating in East Texas, making it challenging to approximate the extent of the original longleaf pine forest in the state. A reasonable estimate is that longleaf pine forests once covered 6 million acres in eastern Texas.

The story of Texas's longleaf pine reads almost identically to that of Louisiana or Mississippi. Investment money came in, and longleaf pine forests were cut out. Long-forgotten sawmill towns like Aldridge, Manning, and Buna shipped railcar after railcar of longleaf lumber to export from Houston, Galveston, and other bustling port cities. The sturdy longleaf pine even played a vital role in (wooden) shipbuilding in coastal Texas cities like the Beaumont–Port Arthur area. The U.S. Forest Service estimates that only 40,000 acres of longleaf pine are now found in Texas, but this is likely an underestimate. Nonetheless, much of what remains in Texas is being actively restored. Some places include the following locations.

Roy E. Larsen Sandyland Sanctuary. Hardin County, Silsbee area. Owned and managed by the Nature Conservancy, this area is open daily for self-guided tours highlighting longleaf pine and a number of rare plant species. Still a work in progress, many of the elements of a functioning longleaf pine forest are in place. There is ample sunlight reaching the forest floor, trees are widely spaced, and regular prescribed burning is taking place. The one element still needed is time. Specifically, the trees are relatively young. That doesn't detract from this forest, which harbors typical plants seen in sandhill ecosystems, including yucca, prickly pear cactus, and blazing star species. Both the Sandhill Loop Trail and Longleaf Loop Trail take you through the nicer longleaf pine forests.

From Silsbee at the intersection of Texas State Highway 327 and U.S. Highway 96, travel 3.4 miles northwest/west on Highway 327 to the parking lot and trailhead (30°20'56.77"N, 94°14'14.58"W). The parking lot is at 4208 Highway 327 West, Silsbee, Tex. 77656.

Angelina National Forest. Angelina and Jasper Counties, Boykin Springs Recreation Area. This has some of the best examples of natural, fire-maintained longleaf pine forests found in East Texas. The 153,179-acre forest lies in Angelina, Nacogdoches, San Augustine, and Jasper Counties, in the heart of East Texas. Longleaf pine is the predominant cover type in the southern portion, while loblolly and shortleaf pine dominate in the rest of the forest. The 32,300-acre Longleaf Ridge Special Area in the southern part of the forest contains the best longleaf pine here. Head to the Boykin Springs Recreation Area (31°03'42.5"N, 94°16'30.0"W) to see a beautiful fire-maintained stand of old longleaf close to the picnic area. You may even see Red-cockaded Woodpeckers here.

From Zavalla, follow Texas Highway 63 east 10.5 miles. Turn right (south) onto Forest Service Road 313 and travel 2.5 miles to the Boykin Springs campground. An area about 3 miles east of this location and off Highway 63 called the Catahoula Barrens is also worth seeing. There the rolling hills of longleaf pine forests begin to develop a savanna or prairie-like appearance. Check in with the ranger district office for more information at 111 Walnut Ridge Road, Zavalla, Tex. 75980, before heading out.

Sabine National Forest. Sabine County, Fox Hunter's Hill area. This 160,656-acre national forest lies near the Texas–Louisiana border. Most of the best remaining longleaf pine is found in the southern portion of the forest. The longleaf pine in a rolling hill setting looks very similar to that in parts of the Kisatchie National Forest in Louisiana. Among the better longleaf pine forests in the Sabine National Forest is in the Fox Hunter's Hill area (compartment 139) and the Stark Tract (compartments 141–42). The Fox Hunter's Hill area is near 31°10'56.76"N, 93°43'43.28"W.

From Hemphill at the intersection of Texas State Highway 87 and Worth Street, travel south on Highway 87 for 15.4 miles. A small turnoff onto Forest Service Road 113 will be on the left side (east). Although this road is likely to be chained, it is a good spot to carefully pull off and park. Walking this road will take you through the heart of this stand. Though there are various age classes of longleaf pine represented here, the oldest trees are approaching 80 years old. Grasses in the understory include little bluestem and Indiangrass while bush clover and blazing star can also be seen, especially when in bloom in the fall. Check with the Sabine District Office at 5050 Highway 21 East, Hemphill, Tex. 75948, for other opportunities to see longleaf pine forests in this national forest.

Longleaf pine and fire on private land in the longleaf pine range.

Acknowledgments

Our original intent with this project was to create an outreach flier loosely based on material we created for the Longleaf Alliance website in the early 2000s. It grew over a decade into this book. First and foremost, we thank our family, friends, and colleagues who tolerated our late nights, countless periods of lament and anxiety, and temper tantrums as we put this together. Next, we thank the University of Georgia Press for supporting our vision and guiding us through the process of publishing this book. Both the University of Alabama Press and the University of North Carolina Press were also helpful early on in the process by helping us achieve a better direction. We were not interested in a heavily cited academic textbook or dichotomous key, but instead wanted a more readable version—more of a field guide—so that we could introduce nonscientists and the next generation of longleaf pine enthusiasts to this unique forest type.

To appreciate the need for a book such as this, we must look back at those who paved the way for a better understanding and appreciation of longleaf pine forests, fire, and critters who use these forests. Some of those forebearers include (in no particular order) Ellen Call Long, Charles Mohr, William Ashe, Charles Sargent, Wilbur Mattoon, Herman Chapman, Caroline Dorman, G. Frederick Schwarz, Philip Wakeley, Smith Greene, Roland Harper, Thomas Croker, William Wahlenberg, Bill Boyer, Leon Neel, Larry Landers, Angus Gholson, Lane Green, Rick Hatten, George Folkerts, Bob Mitchell, Dean Gjerstad, Joan Walker, Kay Kirkman, Rhett Johnson, Wilson Baker, Todd Engstrom, Julie Moore, Sharon Hermann, Craig Guyer, Cecil Frost, John Kush, Kevin Robertson, Bruce Means, Steve Jack, Jim Barnett, Ken Outcalt,

Dale Brockway, Martin Cipollini, Robert Peet, Dale Wade, Janisse Ray, Beth Maynor Finch, Bill Platt, Bob Franklin, Johnny Stowe, Latimore Smith, Susan Carr, Phil Sheridan, and many, many others.

A number of family, friends, and close colleagues were exceptionally helpful in providing solid critique of this book. These reviewers include Rachel McGuire, Mark Hainds, J. J. Bachant-Brown, Mark Bailey, Roger Birkhead, Jim Cox, John Kush, Johnny Stowe, Randy Browning, Zach Prusak, Jimmy Stiles, Michael Ulyshen, and Rhett Johnson. The UGA Press also engaged two anonymous reviewers to look over this publication. Though the longleaf pine circles are small and we know who the reviewers were, they shall remain anonymous with our gratitude for their thorough review. You'll also see in the photo credits that many photographers have donated their work by way of Creative Commons or outright donation of images. Some of the more notable photographers who graciously did not charge us even though we used a large number of their images include Reed Noss, Pierson Hill, Roger Birkhead, Rachel McGuire, and Alan Cressler. We would not have been able to acquire the few dozen pictures that had to be purchased outright without the financial donations of Charley and Susan Tarver, Traci and Buddy Ingleright, Rhett and Kathy Johnson, Ken and Joanna Nichols, Wade Harrison, Jeremiah Cates, and Kevin Carter. Teri Nye assisted in telling this story with her outstanding artwork. Many, many other people could be thanked, and the omission of their names is by no means an indication of their lack of contribution. Thank-you to those people we neglected to identify by name.

This book is really just the tip of the iceberg of the unique plants, animals, and arthropods that are part of this exquisite ecosystem. Hopefully, it inspires the next generation of longleaf pine enthusiasts to get out and explore these piney woods.

The sketch artwork by Teri Nye is based on work supported by the U.S. Department of Agriculture, under agreement number 68-4101-16-1006. Any opinions, findings, conclusions, or recommendations expressed in this publication are those of the authors and do not necessarily reflect the views of the U.S. Department of Agriculture. In addition, any reference to specific brands or types of products or services does not constitute or imply an endorsement by the U.S. Department of Agriculture for those products or services. USDA is an equal opportunity provider and employer.

Species List

This list includes the species and subspecies mentioned in the text. Nonnative and invasive species are marked with an asterisk. This is just a smattering of the life that abounds in longleaf pine forests. The adventure awaits you to find the rest.

Mammals

American bison (*Bison bison*)
Ancient bison (*Bison antiquus*)
Baird's pocket gopher (*Geomys breviceps*)
Black bear (*Ursus americanus*)
Bobcat (*Lynx rufus*)
Carolina yellow dog (*Canis lupus familiaris*)
Cotton mouse (*Peromyscus gossypinus*)
Coyote (*Canis latrans*)
Elk (*Cervus elaphus*)
Eastern gray squirrel (*Sciurus carolinensis*)
Eastern red bat (*Lasiurus borealis*)
Eastern spotted skunk (*Spilogale putorius*)
Evening bat (*Nycticeius humeralis*)
Feral hog (*Sus scrofa*)*
Florida mouse or gopher mouse (*Podomys floridanus*)
Fox squirrel (*Sciurus niger*)
Golden mouse (*Ochrotomys nuttalli*)
Gray fox (*Urocyon cinereoargenteus*)
Hispid cotton rat (*Sigmodon hispidus*)
Least shrew (*Cryptotis parva*)
Little brown bat (*Myotis lucifugus*)
Nine-banded armadillo (*Dasypus novemcinctus*)
Northern long-eared bat (*Myotis septentrionalis*)
Northern yellow bat (*Lasiurus intermedius*)
Oldfield mouse (*Peromyscus polionotus*)
Pine vole (*Microtus pinetorum*)
Pineywoods cattle (*Bos taurus*)
Raccoon (*Procyon lotor*)
Red fox (*Vulpes vulpes*)
Red wolf (*Canis rufus*)
Seminole bat (*Lasiurus seminolus*)
Southeastern myotis (*Myotis austroriparius*)
Southeastern pocket gopher (*Geomys pinetis*)
Southeastern shrew (*Sorex longirostris*)
Southern flying squirrel (*Glaucomys volans*)
Southern short-tailed shrew (*Blarina carolinensis*)
Striped skunk (*Mephitis mephitis*)
Tricolored bat (*Perimyotis subflavus*)
Virginia opossum (*Didelphis virginiana*)
White-tailed deer (*Odocoileus virginianus*)

Reptiles

Banded watersnake (*Nerodia fasciata*)
Black pine snake (*Pituophis melanoleucus lodingi*)
Blue-tailed mole skink (*Plestiodon egregius lividus*)
Box turtle (*Terrapene carolina*)
Broadhead skink (*Plestiodon laticeps*)
Brown anole (*Anolis sagrei*)*
Carolina pigmy rattlesnake (*Sistrurus miliarius miliarius*)
Cedar Key mole skink (*Plestiodon egregius insularis*)
Coal skink (*Plestiodon anthracinus*)
Copperhead (*Agkistrodon contortrix*)
Corn snake (*Pantherophis guttatus*)
Cottonmouth (*Agkistrodon piscivorus*)
Dusky pigmy rattlesnake (*Sistrurus miliarius barbouri*)
Gopher tortoise (*Gopherus polyphemus*)
Gulf Coast box turtle (*Terrapene carolina major*)
Ground skink (*Scincella lateralis*)
Eastern box turtle (*Terrapene carolina carolina*)
Eastern coachwhip (*Masticophis flagellum*)
Eastern coral snake (*Micrurus fulvius*)
Eastern diamondback rattlesnake (*Crotalus adamanteus*)
Eastern fence lizard (*Sceloporus undulatus*)
Eastern glass lizard (*Ophisaurus ventralis*)
Eastern hognose snake (*Heterodon platirhinos*)
Eastern indigo snake (*Drymarchon corais couperi*)
Eastern kingsnake (*Lampropelits getula*)
Eastern kingsnake subspecies (*Lampropeltis getula getula*)
Eastern rat snake (*Pantherophis obsoletus*)
Five-lined skink (*Plestiodon fasciatus*)
Florida box turtle (*Terrapene carolina bauri*)
Florida Keys mole skink (*Plestiodon egregius egregius*)
Florida pine snake (*Pituophis melanoleucus mugitus*)
Gray rat snake (*Pantherophis spiloides*)
Green anole (*Anolis carolinensis*)
Island glass lizard (*Ophisaurus compressus*)
Louisiana pine snake (*Pituophis ruthveni*)
Mimic glass lizard (*Ophisaurus mimicus*)
Mole skink (*Plestiodon egregious*)
Northern black racer (*Coluber constrictor constrictor*)
Northern mole skink (*Plestiodon egregius similis*)
Northern pine snake (*Pituophis melanoleucus melanoleucus*)
Northern watersnake (*Nerodia sipedon*)
Peninsula mole skink (*Plestiodon egregius onocrepis*)
Pigmy rattlesnake (*Sistrurus miliarius*)
Pine snake (*Pituophis melanoleucus*)
Pinewoods snake (*Rhadinaea flavilata*)
Scarlet kingsnake (*Lampropeltis triangulum elapsoides*)
Scarlet snake (*Cemophora coccinea*)
Short-tailed snake (*Lampropeltis extenuata*)
Six-lined racerunner (*Cnemidophorus sexlineatus*)
Slender glass lizard (*Ophisaurus attenuatus*)
Southeastern crowned snake (*Tantilla coronata*)
Southeastern five-lined skink (*Plestiodon inexpectatus*)
Southern black racer (*Coluber constrictor priapus*)
Southern hognose snake (*Heterodon simus*)
Speckled kingsnake (*Lampropeltis getula holbrooki*)
Three-toed box turtle (*Terrapene carolina triunguis*)
Timber rattlesnake (*Crotalus horridus*)
Western pigmy rattlesnake (*Sistrurus miliarius streckeri*)

Amphibians

American toad (*Anaxyrus americanus*)
Atlantic coast leopard frog (*Lithobates kauffeldi*)
Barking treefrog (*Hyla gratiosa*)
Bog dwarf salamander (*Eurycea sphagnicola*)
Brimley's chorus frog (*Pseudacris brimleyi*)
Broken-striped newt (*Notophthalmus viridescens dorsalis*)
Chamberlain's dwarf salamander (*Eurycea chamberlaini*)
Dwarf salamander (*Eurycea quadridigitata*)
Eastern narrowmouth toad (*Gastrophryne carolinensis*)
Eastern spadefoot (*Scaphiopus holbrooki*)
Eastern tiger salamander (*Ambystoma tigrinum*)
Frosted flatwoods salamander (*Ambystoma cingulatum*)
Gopher frog (*Lithobates capito*)
Green treefrog (*Hyla cinerea*)
Hillis's dwarf salamander (*Eurycea hillisi*)
Little grass frog (*Pseudacris ocularis*)
Mississippi gopher frog (*Lithobates sevosus*)
Oak toad (*Anaxyrus quercicus*)
Ornate chorus frog (*Pseudacris ornata*)
Pickerel frog (*Lithobates palustris*)
Pine barrens treefrog (*Hyla andersonii*)
Pine woods treefrog (*Hyla femoralis*)
Red-spotted newt (*Notopthalmus viridescens*)
Reticulated flatwoods salamander (*Ambystoma bishopi*)
Southern chorus frog (*Pseudacris nigrita*)
Southern leopard frog (*Lithobates sphenocephalus*)
Southern toad (*Anaxyrus terrestris*)
Spotted salamander (*Ambystoma maculatum*)
Spring peeper (*Pseudacris crucifer*)
Squirrel treefrog (*Hyla squirella*)
Striped newt (*Notophthalmus perstriatus*)
Western dwarf salamander (*Eurycea paludicola*)

Birds

American Kestrel (*Falco sparverius*)
Bachman's Sparrow (*Peucaea aestivalis*)
Black-billed Cuckoo (*Coccyzus erythropthalmus*)
Black Vulture (*Coragyps atratus*)
Blue Grosbeak (*Passerina caerulea*)
Blue Jay (*Cyanocitta cristata*)
Brown-headed Nuthatch (*Sitta pusilla*)
Carolina Chickadee (*Poecile carolinensis*)
Carolina Wren (*Thryothorus ludovicianus*)
Chipping Sparrow (*Spizella passerina*)
Common Nighthawk (*Chordeiles minor*)
Common Yellowthroat (*Geothlypis trichas*)
Cooper's Hawk (*Accipiter cooperii*)
Downy Woodpecker (*Dryobates pubescens*)
Eastern Bluebird (*Sialia sialis*)
Eastern Kingbird (*Tyrannus tyrannus*)
Eastern Phoebe (*Sayornis phoebe*)
Eastern Screech Owl (*Megascops asio*)
Eastern Towhee (*Pipilo erythrophthalmus*)
Eastern Wild Turkey (*Meleagris gallopavo silvestris*)
Eastern Wood Pewee (*Contopus virens*)
European Starling (*Sturnus vulgaris*)*
Field Sparrow (*Spizella pusilla*)
Gray Catbird (*Dumetella carolinensis*)
Great Crested Flycatcher (*Myiarchus crinitus*)
Great Horned Owl (*Bubo virginianus*)
Ground Dove (*Columbina passerina*)
Hairy Woodpecker (*Dryobates villosus*)
Henslow's Sparrow (*Ammodramus henslowii*)
Hooded Warbler (*Setophaga citrina*)
Indigo Bunting (*Passerina cyanea*)
Loggerhead Shrike (*Lanius ludovicianus*)
Mourning Doves (*Zenaida macroura*)
Northern Bobwhite (*Colinus virginianus*)

Northern Flicker (*Colaptes auratus*)
Northern Mockingbird (*Mimus polyglottos*)
Orange-crowned Warbler (*Leiothlypis celata*)
Orchard Oriole (*Icterus spurius*)
Palm Warbler (*Setophaga palmarum*)
Pileated Woodpecker (*Dryocopus pileatus*)
Pine Warbler (*Setophaga pinus*)
Prairie Warbler (*Setophaga discolor*)
Red-bellied Woodpecker (*Melanerpes carolinus*)
Red-breasted Nuthatch (*Sitta canadensis*)
Red-cockaded Woodpecker (*Picoides borealis*)
Red-headed Woodpecker (*Melanerpes erythrocephalus*)
Red-shouldered Hawk (*Buteo lineatus*)
Red-tailed Hawk (*Buteo jamaicensis*)
Ruby-throated Hummingbird (*Archilochus colubris*)
Song Sparrow (*Melospiza melodia*)
Southeastern Kestrel (*Falco sparverius paulus*)
Summer Tanager (*Piranga rubra*)
Swamp Sparrow (*Melospiza georgiana*)
Tufted Titmouse (*Baeolophus bicolor*)
Turkey Vulture (*Cathartes aura*)
Yellow-bellied Sapsucker (*Sphyrapicus varius*)
Yellow-billed Cuckoo (*Coccyzus americanus*)
Yellow-rumped Warbler (*Setophaga coronata*)
Yellow-throated Warbler (*Setophaga dominica*)
White-breasted Nuthatch (*Sitta carolinensis*)
White-throated Sparrow (*Zonotrichia albicollis*)

Arthropods

Acrobat ant (*Crematogaster ashmeadi*)
American grasshopper (*Schistocerca americana*)
American grass mantis (*Thesprotia graminis*)
Antlion (*Myrmeleon carolinus*)
Black imported fire ants (*Solenopsis richteri*)*
Black-legged tick or deer tick (*Ixodes scapularis*)
Black turpentine beetle (*Dendroctonus terebrans*)
Brown dog tick (*Rhipicephalus sanguineus*)
Cactus coreid (*Chelinidea vittiger aequoris*)
Carpenter ant (*Camponotus nearcticus*)
Cloudless sulfur butterfly (*Phoebis sennae*)
Coarse-writing engraver (*Ips calligraphus*)
Common eastern bumblebee (*Bombus impatiens*)
Common long-horned bee (*Melissodes communis*)
Cone ant (*Dorymyrmex bureni*)
Eastern five-spined engraver (*Ips grandicollis*)
Eastern lubber grasshopper (*Romalea guttata*)
Eastern tent caterpillar (*Malacosoma americana*)
Field ant (*Formica pallidefulva*)
Florida blue centipede (*Scolopendra viridis*)
Florida harvester ant (*Pogonomyrmex badius*)
Florida small carpenter bee (*Certaina floridana*)
Florida woods cockroach (*Eurycotis floridana*)
Frosted elfin butterfly (*Cullophrys irus*)
Fulvous wood roach (*Parcoblatta fulvescens*)
Giant wolf spider (*Hogna carolinensis*)
Goatweed leafwing butterfly (*Anaea andria*)
Golden green sweat bee (*Augochlorella aurata*)
Golden orb weaver spider (*Trichonephila clavipes*)
Gopher cricket (*Ceuthophilus latibuli*)
Gopher tick (*Amblyomma tuberculatum*)
Gopher tortoise aphodius beetle (*Aphodius laevigatus*)
Gopher tortoise copris beetle (*Copris gopher*)
Gopher tortoise hister beetle (*Chelyoxenus xerobatis*)
Gopher tortoise onthophagus beetle (*Onthophagus polyphemi*)
Green lynx spider (*Peucetia viridans*)
Gulf Coast tick (*Amblyomma maculatum*)
Gulf fritillary butterfly (*Agraulis vanillae*)
Hentz striped scorpion (*Centruroides hentzi*)
High noon ant (*Forelius pruinosus*)

Karner blue butterfly (*Plebejus melissa samuelis*)
Large milkweed bug (*Oncopeltus fasciatus*)
Leaf-footed cactus bug (*Leptoglossus phyllopus*)
Little gopher tortoise scarab beetle (*Alloblackburneus troglodytes*)
Lone star tick (*Amblyomma americanum*)
Milkweed assassin bug (*Zelus longipes*)
Morris big-headed ant (*Pheidole morrisi*)
Myrmicine ant (*Temnothorax palustris*)
Palmetto tortoise beetle (*Hemisphaerota cyanea*)
Red-banded hairstreak butterfly (*Calycopis cecrops*)
Red imported fire ants (*Solenopsis invicta*)*
Reticulate metallic sweat bee (*Lasioglossum reticulatum*)
Rugose-chested sweat bee (*Lasioglossum pectorale*)
Sculptured pine borer (*Chalcophora virginiensis*)
Small southern pine engraver (*Ips avulsus*)
Southern pine beetle (*Dendroctonus frontalis*)
Southern pine sawyer (*Monochamus titillator*)
Southern plains bumblebee (*Bombus fraternus*)
Southern two-striped walking stick (*Anisomorpha buprestoides*)
Southern unstriped scorpion (*Vaejovis carolinianus*)
Spicebush swallowtail butterfly (*Papilio troilus*)
Spiny backed orb weaver (*Gasteracantha cancriformis*)
Spotted oak apple gall wasp (*Atrusca quercuscentricola*)
Stinging paper wasps (*Polistes carolina*)
Tawny crazy ant (*Nylanderia fulva*)*
Tiger swallowtail (*Papilio glaucus*)
Thief ant (*Solenopsis carolinensis*)
Turpentine borer (*Buprestis apricans*)
Vampire ant (*Stigmatomma pallipes*)
Velvet ant (*Dasymutilla occidentalis*)
Walker's camel cricket (*Ceuthophilus walkeri*)
Wild indigo duskywing (*Erynnis baptisiae*)

Plants

GRASSES

Arrowfeather three-awn (*Aristida purpurascens*)
Beaked panicgrass (*Panicum anceps*)
Big bluestem (*Andropogon gerardii*)
Blackseed needlegrass (*Piptochaetium avenaceum*)
Broomsedge (*Andropogon virginicus*)
Bushy bluestem (*Andropogon glomeratus*)
Cogongrass (*Imperata cylindrica*)*
Elliott's bluestem (*Andropogon gyrans*)
Florida dropseed (*Sporobolus floridanus*)
Hairawn muhly grass (*Muhlenbergia capillaris*)
Little bluestem (*Schizachyrium scoparium*)
Lopsided Indiangrass (*Sorghastrum secundum*)
Pineywoods dropseed (*Sporobolus junceus*)
Purple lovegrass (*Eragrostis spectabilis*)
River cane (*Arundinaria* spp.)
Savanna hairgrass (*Muhlenbergia expansa*)
Slender bluestem (*Schizachyrium tenerum*)
Slender Indiangrass (*Sorghastrum elliottii*)
Splitbeard bluestem (*Andropogon ternarius*)
Switchgrass (*Panicum virgatum*)
Toothache grass (*Ctenium aromaticum*)
Yellow Indiangrass (*Sorghastrum nutans*)
Wiregrass (*Aristida stricta*)
Wiregrass (*Aristida beyrichiana*)
Witch grasses (*Dicanthelium* spp.)
Woolysheath three-awn (*Aristida lanosa*)

FORBS

Anisescented goldenrod (*Solidago odora*)
Atlantic pigeonwings (*Clitoria mariana*)
Bicolor lespedeza (*Lespedeza bicolor*)*
Black-eyed Susan (*Rudbeckia hirta*)
Bladderworts (*Utricularia* spp.)

Bracken fern (*Pteridium aquilinum*)
Butterfly weed (*Asclepias tuberosa*)
Butterworts (*Pinguicula* spp.)
Canada goldenrod (*Solidago canadensis*)
Carolina silkgrass (*Pityopsis aspera*)
Carolina wild indigo (*Baptisia cinerea*)
Chinese bush clover (*Lespedeza cuneata*)*
Clasping milkweed (*Asclepias amplexicaulis*)
Coastalplain tickseed (*Coreopsis gladiata*)
Common bog button (*Lachnocaulon anceps*)
Coral Bean (*Erythrina herbacea*)
Cottony golden aster (*Chrysopsis gossypina*)
Crested fringed orchid (*Platanthera cristata*)
Dixie white-topped aster (*Sericocarpus tortifolius*)
Dollarleaf (*Rhynchosia reniformis*)
Doveweed (*Croton glandulosus*)
Downy eastern milkpea (*Galactia volubilis*)
Eastern prickly pear (*Opuntia mesacantha*)
Eastern sensitive briar (*Mimosa microphylla*)
Erect milkpea (*Galactia erecta*)
Erect prickly pear (*Opuntia stricta*)
Fewflower milkweed (*Asclepias lanceolata*)
Florida hoary pea (*Tephrosia florida*)
Florida Keys sensitive pea (*Chamaecrista deeringiana*)
Florida ticktrefoil (*Desmodium floridanum*)
Goat's rue (*Tephrosia virginiana*)
Golden aster (*Chrysopsis mariana*)
Gopherweed (*Baptisia lanceolata*)
Grassleaf goldaster (*Pityopsis oligantha*)
Grass pink (*Calopogon tuberosus*)
Hairy lespedeza (*Lespedeza hirta*)
Hairy St. John's wort (*Hypericum setosum*)
Handsome Harry (*Rhexia virginica*)
Hatpin (*Eriocaulon decangulare*)
Healing croton (*Croton argyranthemus*)
Hooded pitcher plant (*Sarracenia minor*)
Justiceweed (*Eupatorium leucolepis*)
Littleleaf ticktrefoil (*Desmodium ciliare*)
Lobed tickseed (*Coreopsis auriculata*)
Maryland meadow beauty (*Rhexia mariana*)
Michaux's milkweed (*Asclepias michauxii*)
Mohr's coneflower (*Rudbeckia mollis*)
Nuttall's wild indigo (*Baptisia nuttalliana*)
Orange coneflower (*Rudbeckia fulgida*)
Orange fringed orchid (*Platanthera ciliaris*)
Orange milkwort (*Polygala lutea*)
Pale grass pink (*Calopogon pallidus*)
Pale pitcher plant (*Sarracenia alata*)
Panhandle lily (*Lilium iridollae*)
Partridge pea (*Chamaecrista fasciculata*)
Pencil flower (*Stylosanthes biflora*)
Pineland golden aster (*Chrysopsis latisquamea*)
Pineland St. John's wort (*Hypericum suffruticosum*)
Pineland ticktrefoil (*Desmodium strictum*)
Pineland wild indigo (*Baptisia lecontei*)
Pine lily (*Lilium catesbaei*)
Pinkscale blazing star (*Liatris elegans*)
Poison ivy (*Toxicodendron radicans*)
Pokeweed (*Phytolacca americana*)
Prairie blazing star (*Liatris pycnostachya*)
Pursh's rattlebox (*Crotalaria purshii*)
Rabbitbell (*Crotalaria rotundifolia*)
Rice button aster (*Symphyotrichum dumosum*)
Rose pogonia (*Pogonia ophioglossoides*)
Roundleaf thoroughwort (*Eupatorium rotundifolium*)
St. Andrew's cross (*Hypericum hypericoides*)
Sandhill dawnflower (*Stylisma patens*)
Sandhill lupine (*Lupinus diffusus*)
Sandhill milkpea (*Galactia regularis*)
Sandhill milkweed (*Asclepias humistrata*)
Sandhill St. John's wort (*Hypericum tenuifolium*)
Sand rushfoil (*Croton michauxii*)
Sand ticktrefoil (*Desmodium lineatum*)

Savanna meadow beauty (*Rhexia alifanus*)
Scaleleaf aster (*Symphyotrichum adnatus*)
Scurf hoary pea (*Tephrosia chrysophylla*)
Shaggy blazing star (*Liatris pilosa*)
Showy rattlebox (*Crotalaria spectabilis*)*
Silkgrass (*Pityopsis graminifolia*)
Silvery aster (*Symphyotrichum concolor*)
Silvery Bush Clover (*Lespedeza capitata*)
Slender scaly blazing star (*Liatris squarrosa*)
Small-flowered partridge pea (*Chamaecrista nictitans*)
Smallflower thoroughwort (*Eupatorium semiserratum*)
Small's bog button (*Lachnocaulon minus*)
Smooth ticktrefoil (*Desmodium laevigatum*)
Soft greeneyes (*Berlandiera pumila*)
Soft milkpea (*Galactia mollis*)
Spiked hoary pea (*Tephrosia spicata*)
Spurred butterfly Pea (*Centrosema virginianum*)
Star tickseed (*Coreopsis pubescens*)
Stiff-leaved aster (*Ionactis linariifolius*)
Sundews (*Drosera* spp.)
Sundial lupine (*Lupinus perennis*)
Swamp milkweed (*Asclepias incarnata*)
Tall ironweed (*Vernonia angustifolia*)
Taylor county silkgrass (*Pityopsis pinifolia*)
Texas bull nettle (*Cnidoscolus texanus*)
Thunberg lespedeza (*Lespedeza thunbergii*)*
Tread softly (*Cnidoscolus stimulosus*)
Trumpet pitcher plant (*Sarracenia flava*)
Yankeeweed (*Eupatorium compositifolium*)
Yellow hatpin (*Syngonanthus flavidulus*)
Yellow meadow beauty (*Rhexia lutea*)
Venus flytrap (*Dionaea muscipula*)
Walter's aster (*Symphyotrichum walteri*)
Wand goldenrod (*Solidago stricta*)
Whitetop pitcher plant (*Sarracenia leucophylla*)
White wild indigo (*Baptisia alba*)
Whorled coreopsis (*Coreopsis major*)
Wooly croton (*Croton capitatus*)
Wrinkleleaf goldenrod (*Solidago rugosa*)
Yucca (*Yucca filamentosa*)

SHRUBS

Allegheny chinkapin (*Castanea pumila*)
American beautyberry (*Callicarpa americana*)
American holly (*Ilex opaca*)
Atlantic poison oak (*Toxicodendron pubescens*)
Bigflower pawpaw (*Asimina obovata*)
Buckberry (*Gaylussacia ursina*)
Carolina holly (*Ilex ambigua*)
Dangleberry (*Gaylussacia frondosa*)
Deerberry (*Vaccinium stamineum*)
Dwarf deerberry (*Vaccinium stamineum* var. *arenicola*)
Dwarf huckleberry (*Gaylussacia dumosa*)
Dwarf pawpaw (*Asimina pygmaea*)
Elliott's blueberry (*Vaccinium elliottii*)
Fragrant sumac (*Rhus aromatica*)
Gopher apple (*Licania michauxii*)
Inkberry (*Ilex glabra*)
Large gallberry (*Ilex coriacea*)
Needle palm (*Rhapidophyllum hystrix*)
Pixie moss (*Pyxidanthera barbulata*)
Poison sumac (*Toxicodendron vernix*)
Running oak (*Quercus elliottii*)
Saw palmetto (*Serenoa repens*)
Shiny blueberry (*Vaccinium myrsinites*)
Slimleaf pawpaw (*Asimina angustifolia*)
Smooth sumac (*Rhus glabra*)
Southern highbush blueberry (*Vaccinium formosum*)
Sparkleberry (*Vaccinium arboreum*)
Titi (*Cyrilla racemiflora*)
Wax myrtle (*Morella cerifera*)
Winged sumac (*Rhus copallinum*)
Yaupon holly (*Ilex vomitoria*)

VINES, CLIMBING SHRUBS, AND BRAMBLES

Carolina jessamine (*Gelsemium sempervirens*)
Coral honeysuckle (*Lonicera sempervirens*)
Japanese climbing fern (*Lygodium japonicum*)*
Lance-leaf greenbrier (*Smilax smallii*)
Laurel leaf greenbrier (*Smilax laurifolia*)
Muscadine (*Vitus rotundifolia*)
Northern dewberry (*Rubus flagellaris*)
Purple passionflower (*Passiflora incarnata*)
Saw greenbrier (*Smilax bona-nox*)
Southern dewberry (*Rubus trivialis*)
Swamp jessamine (*Gelsemium rankinii*)
Virginia creeper vine (*Parthenocissus quinquefolia*)

TREES

American chestnut (*Castanea dentata*)
Blackjack oak (*Quercus marilandica*)
Black tupelo (*Nyssa biflora*)
Bluejack oak (*Quercus incana*)
Cabbage palm (*Sabal palmetto*)
Chinese tallow tree (*Triadica sebifera*)*
Flowering dogwood (*Benthamidia florida*)
Loblolly pine (*Pinus taeda*)
Longleaf pine (*Pinus palustris*)
Mimosa tree (*Albizia julibrissin*)*
Pawpaw (*Asimina triloba*)
Pond cypress (*Taxodium ascendens*)
Sand post oak (*Quercus margaretta*)
Sassafras (*Sassafras albidum*)
Shortleaf pine (*Pinus echinata*)
Slash pine (*Pinus elliottii*)
Sourwood (*Oxydendrum arboreum*)
Turkey oak (*Quercus laevis*)
Wild persimmon (*Diospyros virginiana*)
Wild plum (*Prunus umbellata*)
Willow oak (*Quercus phellos*)

LICHEN, MOSS, AND FUNGUS

Cochineal scale (*Dactylopious* spp.)
Cup lichen (*Cladonia leporina*)
Dixie reindeer lichen (*Cladonia subtenuis*)
Powderpuff lichen (*Cladonia evansii*)
Red heart fungus (*Phellinus pini*)
Reindeer lichen (*Cladina rangiferina*)
White-nose fungus (*Pseudogymnoascus destructans*)

Cryptozoology

Bardin Booger
Splinter cat (*Felynx arbordiffisus*)
Hoop snake (*Serpenscirculousus caudavenenifer*)

Glossary

agroforestry	land use that combines agriculture and forestry; for example, silvopasture that combines cattle grazing and growing trees
annual	plants that finish their life cycle during one growing season (1 year)
apron	the opening of a gopher tortoise burrow
arthropods	a group of animals that includes insects and spiders
autotomy	process by which some reptiles shed portions of their tail to evade predators
awn	a narrow bristlelike attachment on the flowering structure of a grasslike plant
axil	the place on a plant where the leaf or leaf stalk meets the stem
biennial	a plant that lives for two growing seasons, usually in a nonreproductive vegetative state the first season, maturing to an "adult" reproductive stage the following growing season. This can take place in as little as one calendar year.
bract	leaflike structure on a plant, typically as part of the inflorescence
cache	to store away food supplies
candle	the early season growing tip of a longleaf pine that resembles a white candle before needle growth
canopy	the uppermost horizontal structure of a forest, often consisting of tree species
carapace	the top shell of a turtle or tortoise

Carolina bay	an elliptical wetland in a shallow depression that is frequently fed by rain and groundwater
cavity	a hollowed-out chamber excavated by certain species of wildlife in a dead woody structure or one excavated in a living pine tree by Red-cockaded Woodpeckers
commensals	animals that obtain food, refuge, and other benefits from another animal
compound leaf	a leaf that is dissected into multiple leaflets along its rachis and/or veins
coniferous	plants that bear cones
cranial	relating to the skull
cryptic	coloring exhibited on an animal to camouflage it from predators; species that are not easily identifiable by field characteristics alone
defoliation	the loss of leaves
detritus	decaying organic material on the forest floor
dewlap	a fold of skin below the chin on some lizards
disk flower	one of two types of flowers that are seen in the composite flowers of the Asteraceae or sunflower family; a small and tubular flower that forms in aggregate the disk of the composite flower head
dispersal	the process by which seeds are distributed from a plant; for example, by wind or gravity
dorsal	the top side of an object
drupe	a type of plant fruit with an outer skin and a hard internal seed, such as those found in wild plum and American beautyberry
ecosystem	an interconnected community of organisms that coexist within a specific environment
ecotone	the transitional area that serves as a zone of intermediate topographic, hydrologic, and ecological characteristics of two adjacent ecosystems; for example, longleaf pine forest and Carolina bay
eft	the juvenile stage of an amphibian
forage	to search for food

forb nongraminoid herbaceous plants

fossorial describes animals that spend a significant portion of their life underground

genus the taxonomic division that falls below family; for example, *Pinus*

germination the process by which seeds sprout and begin growth

glochids small hairlike spines with barbs

hatchlings animals newly hatched from eggs

herbaceous describes a nonwoody plant

indigenous native to a geographic area

inflorescence the arrangement of flowers in a plant; for example, raceme, spike, or panicle

invertebrate an animal that lacks a backbone

kits young foxes

larvae(al) an immature stage of development for animals that undergo metamorphosis; for example, as in caterpillars

legume plants that are members of the Fabaceae or pea family

lemma a sheath that covers the seed in grasses

ligule a structure that is present at the juncture of the leaf blade and the sheath in certain grass species

loment a type of fruit found in some members of the pea family; each segment contains one seed and breaks apart for seed dispersal

marsupials mammals that give birth and then house their young in a pouch on the underside of the mother

mast usually the vegetative reproductive part of a woody plant that is sometimes eaten by wildlife; the mast can be hard or soft. Plants with hard masts include acorns, chinkapins, and hickory nuts. Blueberries, blackberries, and plums have soft masts.

mast year a year of high seed production in longleaf pine trees

metamorphosis the process of transformation through life stages from egg to adult

migratory	describes animals that travel from one geographic area to another on a seasonal basis
naval stores	the industry that concentrates on the collection and processing of tree resin for the purpose of ship maintenance
neonate	a newly born snake
nitrogen fixation	the process of converting atmospheric nitrogen into a form that can be taken up by plants
nodules	structures found on the roots of legume species that contain nitrogen-fixing bacteria
nymph	a young insect, sometimes a particular stage within metamorphosis
ovate	egg-shaped
ovoviviparity	the process by which female snakes incubate their eggs inside their bodies until the eggs hatch
parotid glands	kidney-shaped organs behind the eye and ear in some frogs and toads
perennial	describes plants that are reproductive for more than two growing seasons
phasmids	stick insects
pioneer species	plants that are first to establish in disturbed areas
plastron	the bottom shell of a turtle or tortoise
pocosin	derived from an Indigenous word meaning "swamp on a hill"; a wetland habitat that occurs in the coastal plain, often densely vegetated and impenetrable with organic peaty soils
predation	the process by which predators hunt and feed on prey
pronotum	the plate area on back behind the head in insects
prostrate	lying flat, as in the growth habit of a plant
rachis	hardened vascular structures that are often referred to as "veins" on a leaf

ray flower one of two types of flowers that are seen in the composite flowers of the Asteraceae or sunflower family; a strap-like flower that forms in aggregate the "petals" of the composite flower head

refugia habitat or place of refuge for animals

resin another term for the sap of a pine tree

rosette a circular formation of leaves at ground level

rostral scale in snakes, an enlarged scale at the end of their nose

salvage the removal of downed timber following a natural disaster such as tornado or hurricane

sap resinous substance that flows within pine trees

semi-fossorial an animal that requires both underground and aboveground habitats to complete its life cycle

sepals in flowering plants, leafy or petallike structures that sit below or surround the petals and other parts of the flower

simple leaf leaves that are not dissected along the rachis or veins

snag a dead tree that is still standing upright; can be hollow on the inside of the trunk and serves as habitat for many species of wildlife

snout the nose of reptiles and amphibians

species the taxonomic division that falls below genus

spikelets the "flowers" in the grasses

subspecies the taxonomic rank that further subdivides species in plants

symbiont multiple organisms functioning as one; for example, lichens

thigmonasty the ability of a plant to respond relatively rapidly to touch

trifoliate a compound leaf consisting of three leaflets

understory the lower layer of a forest structure that consists of mostly herbaceous plant species

vent the opening at the end of the digestive tract in reptiles

ventral the underside of a snake's body

References

J. Aardema, et. al. Amphibians and Reptiles of North Carolina. https://herpsofnc.org/.

Audubon, J. J. 1834. *Ornithological Biography, Or an account of the habits of the Birds of the United States of America, accompanied by descriptions of the objects represented in the work entitled The Birds of America, and interspersed with delineations of American scenery and manners.* Adam and Charles Black, Edinburgh. 588 pp.

———. 1841. *The Birds of America from Drawings Made in the United States.* Vol. 2. J. B. Chevalier, Philadelphia. 358 pp.

Bartram, W. 1940. *The Travels of William Bartram.* Barnes & Noble, New York. 414 pp.

Chapman, H. H. 1932. Is the longleaf type a climax? Ecology 13(4): 328–334.

Clairborne, J. F. H. 1906. A Trip through the Piney Woods. Publications of the Mississippi Historical Society, IX (November 1906): 487–538.

Croker, T. C., Jr. 1987. Longleaf pine: a history of man and a forest. USDA Forest Service, Southern Region. Forestry Report R8-FR7. 37 pp.

Earley, L. S. 2004. *Looking for Longleaf: The Fall and Rise of an American Forest.* University of North Carolina Press, Chapel Hill, N.C. 322 pp.

Engstrom, R. T. 1993. Characteristic birds and mammals of longleaf pine forest. Tall Timbers Fire Ecol. Conf. 18: 127–138.

Evans, M. 1936. Quoted in *Federal Writers' Project: Slave Narrative Project*, Vol. 2, *Arkansas*, Part 2, *Cannon-Evans*. Manuscript/Mixed Material. www.loc.gov/item/mesn022/.

Finch, B., Maynor Young, B., Johnson, R., and Hall, J. C. 2012. *Longleaf, Far as the Eye Can See: A New Vision of North America's Richest Forest.* University of North Carolina Press, Chapel Hill, N.C. 192 pp.

Folkerts, G. W., Deyrup, M. A., and Sisson, D. C. 1993. Arthropods associated with xeric longleaf pine habitats in the southeastern United States: a brief overview. In: Hermann, S. H., editor. Proceedings of the 18th Tall Timbers fire ecology conference. The longleaf pine ecosystem: ecology, restoration and management; 1991 May 30–June 2; Tallahassee, Fla. Tallahassee (Fla.): Tall Timbers Research Station 18: 159–192.

Frost, C. C. 1993. Four centuries of changing landscape patterns in the longleaf pine ecosystem. In: Hermann, S. H., editor. Proceedings of the 18th Tall Timbers fire ecology conference. The longleaf pine ecosystem: ecology, restoration and management; 1991 May 30–June 2; Tallahassee, Fla. Tallahassee (Fla.): Tall Timbers Research Station 18: 17–44.

Guyer, C., and Bailey, M. A. 1993. Amphibians and reptiles on longleaf pine communities. In: Hermann, S. H., editor. Proceedings of the 18th Tall Timbers fire ecology conference. The longleaf pine ecosystem: ecology, restoration and management; 1991 May 30–June 2; Tallahassee, Fla. Tallahassee (Fla.): Tall Timbers Research Station 18: 139–158.

Hopkins, W. 1951. Woods Hogs vs. Pine Logs. U.S. Department of Agriculture Forest Service, Southern Forest Experiment Station. 14 pp.

Mattoon, W. R. 1922. Longleaf pine. Bulletin 1061. U.S. Department of Agriculture, Washington, D.C. 50 pp.

Means, D. B., and Grow, G. 1985. The endangered longleaf pine community. Enfo. 85(4): 1–12.

Muir, J. 1916. *A Thousand-Mile Walk to the Gulf.* Houghton Mifflin, Boston.

Noss, R. F. 1989. Longleaf pine and wiregrass: keystone components of an endangered Ecosystem. Natural Areas Journal 9(4): 211–213.

Peet, R. K., and Allard, D. J. 1993. Longleaf pine vegetation of the southern Atlantic and eastern Gulf coast regions: a preliminary classification. In: Hermann, S. H., editor. Proceedings of the 18th Tall Timbers fire ecology conference. The longleaf pine ecosystem: ecology, restoration and management; 1991 May 30–June 2; Tallahassee, Fla. Tallahassee (Fla.): Tall Timbers Research Station 18: 45–82.

Ray, J. 1999. *Ecology of a Cracker Childhood.* Milkweed Editions, Minneapolis, Minn. 285 pp.

Savanna River Ecology Laboratory, University of Georgia. Amphibians and herps of South Carolina and Georgia. https://srelherp.uga.edu/herps.htm.

Schwarz G. F. 1907. *The Longleaf Pine in Virgin Forest: A Silvical Study.* John Wiley & Sons, New York. 135 pp.

Speake, S. W. 1981. The gopher tortoise burrow community. In: Proceedings of Gopher Tortoise Council. 44–47.
Stoddard, H. L., Sr. 1962. Use of fire in pine forests and game lands of the Deep Southeast. In: Proceedings of the 1st Tall Timbers fire ecology conference. Tall Timbers Research Station, Tallahassee, Fla. 31–42.
Wahlenberg, W. G. 1946. *Longleaf Pine: Its Use, Ecology, Regeneration, Protection, Growth and Management*. Charles Lathrop Pack Forestry Foundation, Washington, D.C. 429 pp.
Wakeley, P. C. 1954. Planting the southern pines. Agric. Monograph 18. U.S. Department of Agriculture, Forest Service, Washington, D.C. 233 pp.
Walker, L. C. and Wiant, H. W., Jr. 1966. Silviculture of Longleaf Pine. Stephen F. Austin State College. School of Forestry. Texas Bulletin No. 11. 105 pp.
Wells, B. W. 1928. Plant communities of the Coastal Plain of North Carolina and their successional relations. Ecology 9(2): 230–242.

Other Recommended References

Chafin, L. G. 2016. *Field Guide to the Wildflowers of Georgia and Surrounding States*. University of Georgia Press, Athens, Ga. 516 pp.
Godfrey, R. K., and Wooten, J. W. 1979. *Aquatic and Wetland Plants of Southeastern United States—Monocotyledons*. University of Georgia Press, Athens, Ga. 712 pp.
Kaeser, M. J., and Kirkman, L. K. N.d. *Field and Restoration Guide to Common Native Warm-Season Grasses of the Longleaf Pine Ecosystem*. J. W. Jones Ecological Research Center, Newton, Ga. 71 pp.
Miller, J. H., and Miller, K. V. 2005. *Forest Plants of the Southeast and their Wildlife Uses*. University of Georgia Press, Athens, Ga. 454 pp.
Norden, H., and Kirkman, L. K. N.d. *Field Guide to Common Legume Species of the Longleaf Pine Ecosystem*. J. W. Jones Ecological Research Center, Newton, Ga. 72 pp.
Radford, A. E., Ahles, H. E., and Bell, C. R. 1968. *Manual of the Vascular Flora of the Carolinas*. University of North Carolina Press, Chapel Hill, N.C. 1,183 pp.
Sorrie, B. A. 2011. *A Field Guide to Wildflowers of the Sandhills Region*. University of North Carolina Press, Chapel Hill, N.C. 378 pp.

Image Credits

T = top; B = bottom; M = middle; L = left; R = right

ii–iii	Greg Seamon—Tall Timbers
v	Teri Nye
vi	Teri Nye
xii	Alan Cressler
2	Teri Nye
3	Kevin Robertson—Tall Timbers
5	Teri Nye
9	John McGuire
10	Ryan Bollinger
11	Rachel McGuire
12	Justin Meissen licensed under CC BY-SA 2.0
13	Michael D. Martin
14	Martin Cipollini
15	John McGuire
17	Beth Maynor Finch / Longleaf Far as the Eye Can See
18–19	Beth Maynor Finch / Longleaf Far as the Eye Can See
20	Carol Denhof
21	C. Houder licensed under CC BY-NC-ND 2.0
22–23	Carol Denhof
24–25	Teri Nye
26	Rachel McGuire
28	Brian Wiebler—Tall Timbers
29	Toffuti break licensed under CC BY-NC-SA 2.0
31T	Photo Courtesy of Tufts Archives, Pinehurst, N.C.

31B	South Carolina Forestry Commission
32	National Archives, College Park, Ga.
33	Great Southern Lumber Company's Collection, Mss. 3225, Louisiana and Lower Mississippi Valley Collections, LSU Libraries, Baton Rouge, La.
34 both	American Forest Foundation
35	Library of Congress LC-DIG-det-4a20763
36T	American Forestry Association
36B	South Carolina Forestry Commission
37	National Archives, College Park, Ga.
38T	The University of Alabama Libraries Special Collections
38B	Courtesy of Columbus State University Archives, Columbus State University, Columbus, Georgia.
39T	Carol Denhof
39B	William D. Boyer, USDA Forest Service
40	Teri Nye
41	Teri Nye
43L	Kory Roberts under license CC BY-NC
43TR	Paul R. Moosman Jr.
43BR	kmccullough under license CC BY-NC
44	caseytunia licensed under CC BY-NC
46TL	Francois Michonneau licensed under CC BY 4.0
46TR	Joshua Doby licensed under CC BY 4.0
46BL	northganaturalist licensed under CC BY 4.0
46BR	Bill Chitty licensed under CC BY 2.0
47TL	John and Kendra Abbott/Abbott Nature Photography
47TR	Judy Gallagher under license CC BY 2.0
47B	Florida Fish and Wildlife Commission under license CC BY-ND 2.0
48L&TR	Mary Frazer
48MR	Kathleen Smith, Florida Fish and Wildlife Conservation Commission under license CC BY-ND 2.0
48BR	Alan Cressler
49	USFWS
50	Roger Birkhead
52	Florida Fish and Wildlife Commission licensed under CC BY-ND 2.0
53	Todd Belanger licensed under CC BY-NC 4.0
54	Patrick Delaney Florida FWC licensed under CC BY-ND 2.0
55	Teri Nye

56 J. J. Maughn licensed under CC BY-NC-ND 2.0
58 Kara Jones licensed under CC BY-NC 2.0
59L Lee Marcus
59R Jen Cross, USFWS
60 Rich Anderson licensed under CC BY-NC-SA 2.0
62 Florida Fish and Wildlife Commission licensed under CC BY-ND 2.0
63 Gary Cramer USFWS
65L Kenneth Cole Schneider licensed under CC BY-NC-ND 2.0
65R Fyn Kynd licensed under CC BY 2.0
66 Clint Turnage, USFWS
68 Shelly Prevost licensed under CC BY-NC 2.0
69 Coert Dubois in Fearsome Creatures of the Lumberwoods by William T. Cox
70 USFWS Midwest Region
71L Kaylinn Gilstrap
71R Seth Bynum USFWS Southeast Region
72L Kentish Plumber licensed under CC BY-NC-ND 2.0
72R Florida Fish and Wildlife licensed under CC BY-ND 2.0
74 Andy Reago and Chrissy McClarren licensed under CC BY 2.0
75 Andy Reago and Chrissy McClarren licensed under CC BY 2.0
76L vladeb licensed under CC BY-ND 2.0
76R Kenneth Cole Schneider licensed under CC BY-NC-ND 2.0
77 Bill Dickinson licensed under CC BY 2.0
78 Reed Noss
79 Reed Noss
80 Wally Hartshorn licensed under CC BY-ND 2.0
82TL Danny Bales
82TR Laurie Sheppard, U.S. Fish & Wildlife Service Southwest Region licensed under CC PDM 1.0
82B Andy Reago and Chrissy McClarren licensed under CC BY 2.0
83 Andy Reago and Chrissy McClarren licensed under CC BY 2.0
84L Jason Paluck licensed under CC BY-NC-SA 2.0
84R Dapuglet licensed under CC BY-SA 2.0
85 Brady Beck (TLA)
86 Teri Nye
87 Jack Rogers Florida Fish and Wildlife licensed under CC BY-ND 2.0
88 Andy Reago and Chrissy McClarren licensed under CC BY 2.0

89	Matt Stratmoen licensed under CC BY-NC-ND 2.0
90L	YoungSue Public Domain Mark 1.0
90R	Don Verser licensed under CC BY-NC 4.0
91	Andy Reago and Chrissy McClarren licensed under CC BY 2.0
92	Andy Reago and Chrissy McClarren licensed under CC BY 2.0
93	U. S. Fish and Wildlife Service—Northeast Region licensed under CC PDM 1.0
95	Danny Bales
96	Andy Reago and Chrissy McClarren licensed under CC BY 2.0
97	Reed Noss
98	Andy Reago and Chrissy McClarren licensed under CC BY 2.0
100	Danny Bales
101	Danny Bales
102L	Danny Bales
102R	Rick from Alabama licensed under CC BY 2.0
104 both	Andy Reago and Chrissy McClarren licensed under CC BY 2.0
105L	Shenandoah NPS licensed under CC PDM 1.0
105R	Under the Same Moon licensed under CC by 2.0
106L	Skip Russell licensed under CC BY-NC-ND 2.0
106M	Alan Schmierer Public Domain licensed under CC 1.0
106R	Kenneth Cole Schneider licensed under CC BY-NC-ND 2.0
107	Jonathan Bolton
109L	Scott Wahlberg licensed under CC BY-NC
109TR	Kevin Narum
109BR	Byron M. Levan
110	Roger Birkhead
111L	Florida Fish and Wildlife licensed under CC BY-ND 2.0
111R	Pierson Hill
112	Pierson Hill
113	Pierson Hill
114	Pierson Hill
115	Kevin Narum
117	Bradley O'Hanlon Florida FWC licensed under CC BY-ND 2.0
118	Byron M. Levan
120 both	Pierson Hill
121	Byron M. Levan
122	Pierson Hill

123T	Byron M. Levan
123B	Pierson Hill
124	Pierson Hill
126	Byron M. Levan
127	Byron M. Levan
128	Byron M. Levan
129	Pierson Hill
130	Pierson Hill
131TL, BL&R	Pierson Hill
131TR	Kory Roberts licensed under CC BY-NC
133	Pierson Hill
134	Roger Birkhead
135	Pierson Hill
136	Byron M. Levan
137	Byron M. Levan
138	Rachel McGuire
140	Mark Bailey
141 both	Pierson Hill
143	Pierson Hill
144	Pierson Hill
145T&B	Pierson Hill
145M	Byron M. Levan
146	Florida FWC licensed under CC BY-ND 2.0
147	Pierson Hill
148	Pierson Hill
149L	Pierson Hill
149R	Roger Birkhead
150	Pierson Hill
151	Byron M. Levan
152	Pierson Hill
153T	Pierson Hill
153B	Byron M. Levan
154	Pierson Hill
155	Pierson Hill
156	Teri Nye
157	Byron M. Levan
158	Byron M. Levan

160	Teri Nye
161	Teri Nye
162	Brian Gratwicke licensed under CC BY 2.0
163 all	USGS Bee Inventory and Monitoring Lab
165	Teri Nye
166 both	Roger Birkhead
167T	John and Kendra Abbott / Abbott Nature Photography
167B	Rachel McGuire
168	John and Kendra Abbott / Abbott Nature Photography
169L	Alex Wild Photography
169R	Alan Cressler
170 both	Alex Wild Photography
171	Alex Wild Photography
172	Byron M. Levan
173T	Shane Kemp licensed under CC BY-NC-ND 2.0
173B	Jesse Rorabaugh
174	Jonathan Mays
175L	Cletus Lee licensed under CC BY-NC-ND
175R	Tracey Prothro licensed under CC BY-NC
176	Mary Kiem licensed under CC BY-NC-SA 2.0
177	Alan Cressler
178	Judy Gallagher licensed under CC BY 2.0
179	Judy Gallagher licensed under CC BY 2.0
180L	Sequoia Wrens licensed under CC BY-NC
180R	Steve Kerr licensed under CC BY 4.0
181T	Alex Wild Photography
181B	Judy Gallagher licensed under CC BY 2.0
182L	Larah McElroy licensed under CC BY-NC 2.0
182R	Florida Fish and Wildlife Commission licensed under CC BY-ND 2.0
183L	Alex Wild Photography
183R	Judy Gallagher licensed under CC BY 2.0
184L	Christina Butler licensed under CC BY 2.0
184R	Alan Cressler
186	Carol Denhof
187	Carol Denhof
188T	Reed Noss
188B	Macleay Grass Man licensed under CC BY 2.0

189 all	Tom Potterfield licensed under CC BY-NC-SA 2.0
190L	Reed Noss
190R	Carol Denhof
191	dogtooth77 licensed under CC BY-NC-SA 2.0
192	Carol Denhof
193	Sonnia Hill licensed under CC BY 2.0
194	Reed Noss
195	Carol Denhof
196	Larry Allain, U.S. Geological Survey
197	Carol Denhof
198L	Amy Buthod licensed under CC BY-NC-SA 2.0
198R	Mary Keim licensed under CC BY-NC-SA 2.0
200L	Reed Noss
200R	eleanord43 licensed under CC BY-NC 2.0
201	Reed Noss
202L	Byron M. Levan
202R	Reed Noss
203	Reed Noss
204	Carol Denhof
205	Carol Denhof
206 both	Carol Denhof
207	Reed Noss
208	Teri Nye
210L&BR	Carol Denhof
210TR	Scott Zona licensed under CC BY-NC 2.0
211 both	Carol Denhof
212 both	Carol Denhof
213	Carol Denhof
214	Reed Noss
215 both	Carol Denhof
216	Carol Denhof
217 both	Carol Denhof
218	Carol Denhof
219	Carol Denhof
220T	Carol Denhof
220B	Byron M. Levan
221L	Reed Noss

221R	Carol Denhof
222	Laura Clark licensed under CC BY 4.0
223 both	Carol Denhof
224	Reed Noss
225	Carol Denhof
226L	Mary Keim licensed under CC BY-NC-SA 2.0
226R	Carol Denhof
227	Reed Noss
228L	Carol Denhof
228R	Reed Noss
230 both	Carol Denhof
231	dogtooth77 licensed under CC BY-NC-SA 2.0
232 both	Carol Denhof
233	Mary Keim licensed under CC BY-NC-SA 2.0
234	Reed Noss
235T	Carol Denhof
235B	Teri Nye
236 both	Carol Denhof
237	Carol Denhof
239	Reed Noss
240	Carol Denhof
241	Courtesy Alan Cressler, Lady Bird Johnson Wildflower Center
242 both	Carol Denhof
243	Carol Denhof
244	Reed Noss
245	Carol Denhof
246	Adam Arendell licensed under CC BY-NC 2.0
247	Carol Denhof
248	Carol Denhof
249	Mary Keim licensed under CC BY-NC-SA 2.0
250	Carol Denhof
251	Fritz Flohr Reynolds licensed under CC BY-SA 2.0
252 both	Christopher Evans
253	Scott Zona licensed under CC BY-NC 2.0
254	Reed Noss
255	Carol Denhof
256	Mary Keim licensed under CC BY-NC-SA 2.0

258	The Longleaf Alliance
259L	Courtesy Joseph A. Marcus, Lady Bird Johnson Wildflower Center
259R	jessishaw licensed under CC BY-NC
260	Mary Keim licensed under CC BY-NC-SA 2.0
261L	Carol Denhof
261TR	Rachel McGuire
261BR	Nancy Lowenstein
262L	Nancy Loewenstein
262R	Christopher Evans
263	James Gaither licensed under CC BY-NC-ND 2.0
264 all	Rachel McGuire
265L	Christopher Evans
265M	Bruce Kirchoff licensed under CC BY 2.0
265R	Rachel McGuire
266L&M	Nancy Lowenstein
266R	Rachel McGuire
267L&M	Rachel McGuire
267R	Bruce Kirchoff licensed under CC BY 2.0
269 both	Rachel McGuire
270L&R	Rachel McGuire
270M	Mary Keim licensed under CC BY-NC-SA 2.0
271TL&R, BR	Giff Beaton licensed under CC BY-NC
271BL	Tom Walker licensed under CC BY-NC
272L	Keita Watanabe licensed under CC BY-NC
272R	John McGuire
273L	The Longleaf Alliance
273R	Harrison 314 licensed under CC BY-NC
274	USDA Forest Service
275L	Judy Gallagher licensed under CC BY 2.0
275R	Alabama Extension licensed under CC 1.0
277	Carol Denhof
279	Library of Congress
283	Carol Denhof
290	Kevin Robertson
300	Carol Denhof
304–5	Carol Denhof
309	John McGuire

Index